EIK

Elektronische Informationsverarbeitung und Kybernetik

HEFT 1/2
BAND 13
1977

AKADEMIE-VERLAG · BERLIN

REDAKTIONELLE HINWEISE

Manuskripte sind in zwei Exemplaren zu richten an:

Prof. Dr. H. Thiele
Sektion Mathematik der Humboldt-Universität zu Berlin
DDR-1086 Berlin, Postfach 1297

oder

Prof. Dr. W. Kämmerer
DDR-69 Jena, Sonnenbergstr. 6

oder

Dr. H. D. Modrow
Redaktion „EIK" im Zentralinstitut für Kybernetik und Informationsprozesse
DDR-1199 Berlin-Adlershof, Rudower Chaussee 5

Manuskripte werden in deutscher, englischer, französischer und russischer Sprache angenommen.

Die Autoren werden gebeten:

Manuskripte einseitig in Maschinenschrift mit mindestens zweizeiligem Abstand einzureichen (die Zahl der Abbildungen sollte möglichst klein sein).

Namen und Vornamen des Verfassers, seine Anschrift und gegebenenfalls die Institution, in der er tätig ist, sowie deren Leiter anzugeben.

Eine kurze Zusammenfassung der Arbeit von höchstens 10 Zeilen, wenn möglich auch in zwei weiteren der obengenannten Sprachen, beizufügen.

Abbildungen auf besonderen Blättern (Transparentpapier) beizufügen.

Abbildungsunterschriften auf einem gesonderten Blatt beizufügen und, falls bezüglich der Abbildungsgrößen besondere Wünsche bestehen, diese auf den Abbildungen zu vermerken.

In einem Exemplar hervorzuhebende Textstellen schwarz zu unterstreichen, und zwar halbfetter Satz mit einer geraden Linie; kursiv (Schrägschrift) mit einer Wellenlinie; Sperrungen gestrichelt. Besondere Buchstaben durch farbige Unterstreichung zu kennzeichnen; und zwar Fraktur ($\mathfrak{A}$, $\mathfrak{B}$, ...) grün; Griechisch rot; Grotesk (A, B, ...) violett; Schreibschrift ($\mathcal{A}$, $\mathcal{B}$, ...) braun.

Die Autoren erhalten von der Arbeit zwei Korrekturabzüge übersandt. Sie erhalten von ihren Beiträgen 75 Sonderdrucke kostenlos, darüber hinaus bis zu 250 Sonderdrucke gegen Berechnung. Der Verlag behält sich für alle Beiträge das Recht der Vervielfältigung, Verbreitung und Übersetzung innerhalb der gesetzlichen Schutzfrist vor.

Bezugsmöglichkeiten:

Bestellungen sind zu richten

– *in der DDR* an eine Buchhandlung oder
an den Akademie-Verlag,
108 Berlin, Leipziger Straße 3 – 4

– *im sozialistischen Ausland* an eine Buchhandlung für fremdsprachige Literatur oder an den zuständigen Postzeitungsvertrieb

– *in der BRD und in Westberlin* an eine Buchhandlung oder an die Auslieferungsstelle
KUNST UND WISSEN, Erich Bieber,
Wilhelmstraße 4 – 6, 7 Stuttgart 1

– *in Österreich* an den Globus-Buchvertrieb,
1201 Wien, Höchstädtplatz 3

– *im übrigen Ausland* an den Internationalen Buch- und Zeitschriftenhandel; den Buchexport, Volkseigener Außenhandelsbetrieb der Deutschen Demokratischen Republik,
DDR-701 Leipzig, Postfach 160, oder

an den Akademie-Verlag,
DDR-108 Berlin, Leipziger Straße 3 – 4.

HEFT 1/2
BAND 13
SEITE 1–106
JANUAR/FEBRUAR 1977

EIK

Elektronische Informationsverarbeitung und Kybernetik

Herausgegeben im Auftrag
der Mathematischen Gesellschaft der Deutschen Demokratischen Republik
und des Zentralinstituts für Kybernetik und Informationsprozesse
der Akademie der Wissenschaften der DDR
von

WILHELM KÄMMERER (Jena)
und
HELMUT THIELE (Berlin)

AKADEMIE-VERLAG
BERLIN

INHALT

In dieser Zeitschrift werden Beiträge aus folgenden Gebieten und Problemkreisen publiziert: Theorie der Digitalrechner (allgemeine Algorithmentheorie, abstrakte Automaten, algorithmische Sprachen, Programmierung, Struktur konkreter Digitalrechner, Bauelemente); Theorie der Analog- und Hybrid-Rechner; Fragen des praktischen Einsatzes von Rechenautomaten; Schaltstrukturen (kombinatorische Schaltungen, sequentielle Schaltungen, Neuronennetze); wahrscheinlichkeitstheoretische und statistische Probleme (Informationstheorie, Bedienungstheorie, stochastische Automaten, Lernstrukturen, lernende Automaten, Zuverlässigkeitstheorie); Spieltheorie und Theorie der Optimierung; mathematische Linguistik; Anwendungen der elektronischen Informationsverarbeitung und Kybernetik in Natur- und Gesellschaftswissenschaften, soweit sie vorwiegend mathematischen Charakter tragen. – Neben Originalarbeiten werden auch zusammenfassende Berichte über den erreichten Stand der genannten Sachgebiete sowie über einschlägige Tagungen veröffentlicht. Ferner sollen grundlegende Bücher und Einzelarbeiten besprochen werden.

ASMW

Herausgeber: Im Auftrage der Mathematischen Gesellschaft der DDR und des Zentralinstituts für Kybernetik und Informationsprozesse der AdW herausgegeben von Prof. Dr. W. Kämmerer und Prof. Dr. H. Thiele. Verlag: Akademie-Verlag, DDR-108 Berlin, Leipziger Str. 3–4; Fernruf 2200441; Telex-Nr. 114420; Postscheckkonto: Berlin 35021; Bank: Staatsbank der DDR, Berlin, Kto.-Nr.: 6836-26-20712. Chefredaktion: Prof. Dr. Wilhelm Kämmerer, 69 Jena, Sonnenbergstr. 6; Prof. Dr. Helmut Thiele, Sektion Mathematik der Humboldt-Universität zu Berlin, 1086 Berlin, Postfach 1297. Verantwortlicher Redakteur: Dr. Heinz D. Modrow. Anschrift der Redaktion: Zentralinstitut für Kybernetik und Informationsprozesse der AdW, DDR-1199 Berlin, Rudower Chaussee 5; Fernruf: 6702841 App. 2606. Veröffentlicht unter der Lizenznummer 1520 des Presseamtes beim Vorsitzenden des Ministerrates der Deutschen Demokratischen Republik. Gesamtherstellung: VEB Druckerei „Thomas Müntzer", 582 Bad Langensalza. Erscheinungsweise: Die Zeitschrift „Elektronische Informationsverarbeitung und Kybernetik (EIK)" erscheint jährlich in einem Band mit 12 Heften. Bezugspreis je Band 150,– M zuzüglich Versandspesen (Preis für die DDR 120,– M); Preis je Heft 12,50 M (Preis für die DDR 10,– M). Bestellnummer dieses Heftes 1076/13/1–2.

Elektronische Informationsverarbeitung und Kybernetik
EIK 13 (1977) 1/2, 3–12

On Circular Words and ($\omega^* + \omega$)-Powers of Words

By *Keijo Ruohonen*

1. Circular words and ($\omega^* + \omega$)-powers of words introduced

Let A be a finite alphabet. Define in A^* the binary relation Ω by

$$P\Omega Q \quad \text{iff} \quad P = P_1P_2 \quad \text{and} \quad Q = P_2P_1 \text{ for some words } P_1 \text{ and } P_2.$$

Clearly Ω is an equivalence relation with infinite index. The equivalence classes of Ω are called *circular words* over A. An equivalence class of Ω represented by P is denoted by P^Ω. Intuitively, a class $(a_1 \cdots a_n)^\Omega$ where $a_1, \ldots, a_n$ are in A can be identified with the "circular sequence" of Fig. 1. We denote

$$A^{(*)} = \{P^\Omega \mid P \text{ is a word over } A\}.$$

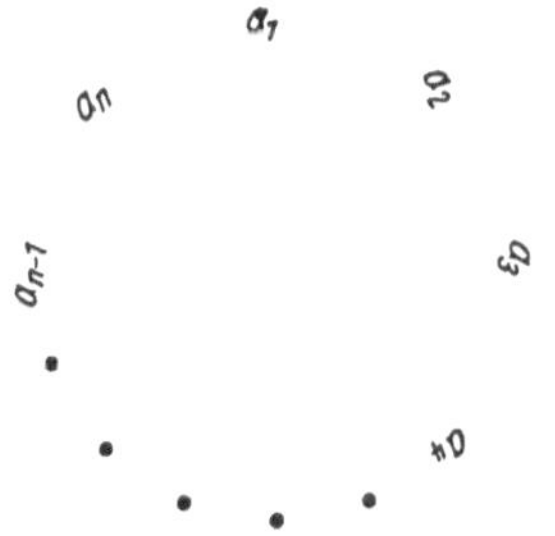

Fig. 1

The *length* of a circular word P^Ω is defined to be the length of P (clearly Ω is included in the relation induced by the length mapping) and is denoted by $|P^\Omega|$, i.e. $|P^\Omega| = |P|$. Although Ω is not a congruence we may note that it is closed under certain unary word operations:

(i) The *mirror image* of a circular word $Q = P^\Omega$ is defined to be $(\text{mi}(P))^\Omega$.

(ii) For each morphism $h: A^* \to B^*$ we define the mapping $h^\Omega: A^{(*)} \to B^{(*)}$ by

$$P^\Omega h^\Omega = (Ph)^\Omega.$$

(iii) The nth *power* of a circular word $Q = P^\Omega$ is defined as $(P^n)^\Omega$ and is denoted by Q^n.

Define in $A^{(*)}$ the binary relation Σ' by

$$P\Sigma'Q \quad \text{iff} \quad P^n = Q^m \text{ for some } n, m > 0.$$

Clearly Σ' is an equivalence relation and so is the relation Σ in A^* given by

$$P\Sigma Q \quad \text{iff} \quad P^\Omega \Sigma' Q^\Omega.$$

The equivalence class of Σ represented by P can be identified with the "doubly infinite" power

$$P^{\omega^*+\omega} = \ldots PPP \ldots$$

For this reason the equivalence classes of Σ are called $(\omega^* + \omega)$-*powers* over A. The class of Σ (resp. Σ') represented by P is denoted by P^Σ (resp. $P^{\Sigma'}$).

Since the unary operations in (i) and (ii) above commute with exponentiation, Σ and Σ' are closed under them. The *length* of an $(\omega^* + \omega)$-power P is

$$|P| = \min \{|Q| \mid Q^\Sigma = P\} .$$

Define the *exponent* $\exp(P)$ and the *root* $\sqrt{P}$ of a nonempty word ("usual" or circular) by

$$\exp(P) = \max \{i \mid P = Q^i \text{ for some word } Q\} ,$$
$$(\sqrt{P})^{\exp(P)} = P .$$

With the help of the following lemmas we prove a basic theorem concerning $(\omega^* + \omega)$-powers.

Lemma 1. *For each nonempty word P, if $P = Q^i$ then i divides* $\exp(P)$.

Proof. Assume, contrary to the claim, that $P = Q^i$ and $i \nmid \exp(P)$. Then $Q = (\sqrt{P})^q R$ for some $q > 0$ and nonempty R of length less than $|\sqrt{P}|$. Since

$$i\,|Q| = i\,|\sqrt{P}|\,q + i\,|R| = |P| = |\sqrt{P}| \exp(P) ,$$

we have

$$|R| = \frac{\exp(P) - iq}{i} |\sqrt{P}| .$$

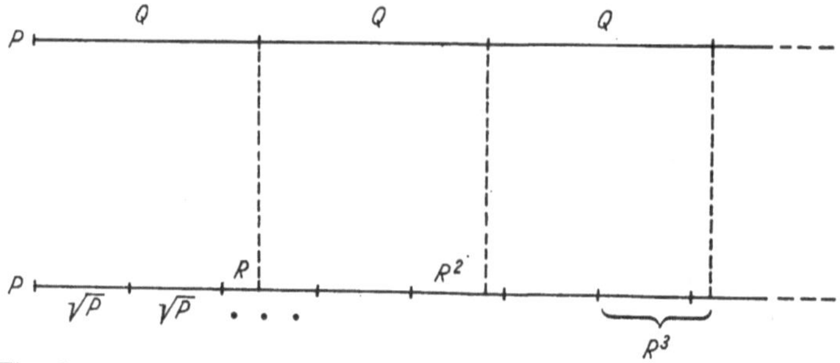

Fig. 2

We find now (see Fig. 2) that

$$R^i = (\sqrt{P})^{\exp(P)-iq} .$$

Proceeding by induction we may assume that the lemma holds true for all nonempty words S of length less than $|P|$. Choosing $S = R^i$ we see that $\sqrt{P} = T^k$ for some word T and $k > 1$ (note that $\exp(P) - iq < i$). This is a contradiction. ∎

Lemma 2. *For each nonempty circular word P, if $P = Q^i$ then i divides* $\exp(P)$.

Proof. For each nonempty word R, $\exp(R) = \exp(R^\Omega)$ and $\sqrt{R^\Omega} = (\sqrt{R})^\Omega$. Thus the lemma follows from Lemma 1. ∎

Theorem 1. *For all nonempty words P and Q, $P^\Sigma = Q^\Sigma$ if and only if $\sqrt{P^\Omega} = \sqrt{Q^\Omega}$.*

Proof. By definition, $P^\Sigma = Q^\Sigma$ if and only if $(P^\Omega)^n = (Q^\Omega)^m$ for some $n, m > 0$. By Lemma 2, this condition is equivalent to $\sqrt{P^\Omega} = \sqrt{Q^\Omega}$. ∎

2. Number of circular words and $(\omega^* + \omega)$-powers of given length

Denote by γ_k the number of circular words in $A^{(*)}$ of length k (for $k \geqq 0$). Further denote

$$C_{k,i} = \{P \in A^* \mid |P| = k, \exp(P) = i\}$$

(for $k, i > 0$).

Theorem 2. *It holds* $\gamma_k = \sum_{i=1}^{k} (i/k) \# C_{k,i}$ *(for* $k > 0$*).*

Proof. Clearly the circular words P^Ω over A with length k and exponent i are exactly those with P in $C_{k,i}$. We claim that

$$\#(C_{k,i} \cap P^\Omega) = (k/i) \qquad (\text{for } P \in C_{k,i}) .$$

The claim which obviously proves the theorem follows when we show that, for any P in $C_{m,1}$, $\# P^\Omega = m$. Assume the contrary, that is $P_1P_2 = P_2P_1 \in C_{m,1}$ for some nonempty words P_1 and P_2. Then

$$\begin{aligned}(P_1P_2)^\Sigma &= (P_1P_2)^{\omega^*} (P_1P_2)^\omega = (P_1P_2)^{\omega^*} P_1(P_2P_1)^\omega \\ &= (P_2P_1)^{\omega^*} P_1(P_1P_2)^\omega = (P_1P_2)^{\omega^*} P_1^2(P_1P_2)^\omega \\ &= \cdots = (P_1P_2)^{\omega^*} P_1^j(P_1P_2)^\omega = \cdots\end{aligned}$$

(for $j \geqq 0$) and so $(P_1P_2)^\Sigma = P_1^\Sigma$. It follows from Theorem 1 that $\sqrt{(P_1P_2)^\Omega} = \sqrt{P_1^\Omega}$. This is however impossible since $\exp((P_1P_2)^\Omega) = \exp(P_1P_2) = 1$. ∎

Corollary 1. *For a prime p or $p = 1$ it holds*

$$\gamma_p = \frac{\# A}{p} (\# A^{p-1} + p - 1) .$$

Proof. The case $p = 1$ is clear. If p is a prime, then

$$C_{p,p} = \{x^p \mid x \text{ is in } A\} , \qquad C_{p,1} = A^p - C_{p,p}$$

and $C_{p,j}$ is empty for $1 < j < p$. ∎

For a nonprime $k > 1$, the formula for γ_k is more complicated. By Theorem 2, it suffices to derive a formula for $\# C_{k,i}$. Furthermore, by the following lemma, it suffices to derive a formula for $\# C_{k,1}$.

Lemma 3. *It holds*

$$\# C_{k,i} = \begin{cases} 0 , & \text{if } i \nmid k \\ \# C_{k/i,1} , & \text{if } i \mid k \end{cases} \qquad (\text{for } k, i > 0) .$$

Proof. Clearly $C_{k,i}$ is empty if $i \nmid k$. If i divides k and $Q^i \in C_{k,i}$, then obviously $\exp(Q) = 1$. On the other hand, if $\exp(Q) = 1$ and $|Q| = k/i$, then Q^i is in $C_{k,i}$. Indeed, if $\exp(Q^i) = j > i$ then, by Lemma 1, i divides j and $Q = (\sqrt{Q^i})^{j/i}$, a contradiction. ∎

Lemma 4. *It holds*

$$\sum_{m|k} \# C_{m,1} = \# A^k \qquad (\text{for } k > 0) .$$

Proof. Clearly

$$\bigcup_{m|k} \{P \in A^k \mid |\sqrt{P}| = m\} = A^k$$

and the union is disjoint. Also clearly

$$\# \{P \in A^k \mid |\sqrt{P}| = m\} = \# C_{m,1} . \blacksquare$$

Denote

$$\sigma_{ij} = \begin{cases} 1 , & \text{if } i \mid j \\ 0 , & \text{if } i \nmid j \end{cases} \qquad (\text{for } i, j > 0)$$

and

$$S_k = \begin{pmatrix} \sigma_{11} \cdots \sigma_{k1} \\ \vdots \quad\quad \vdots \\ \sigma_{1k} \cdots \sigma_{kk} \end{pmatrix} \qquad (\text{for } k > 0) .$$

Then, by Lemma 4, we have

$$S_k \begin{pmatrix} \# C_{1,1} \\ \vdots \\ \# C_{k,1} \end{pmatrix} = \begin{pmatrix} \# A \\ \vdots \\ \# A^k \end{pmatrix} .$$

Let E_k be a matrix such that

$$E_k S_k = \begin{bmatrix} \sigma_{11} & \sigma_{21} & \cdots & \sigma_{k1} \\ \vdots & \vdots & & \vdots \\ \sigma_{1,k-1} & \sigma_{2,k-1} & \cdots & \sigma_{k,k-1} \\ \sigma_{1k} & \frac{1}{2}\sigma_{2k} & \cdots & \frac{1}{k}\sigma_{kk} \end{bmatrix}$$

(note that there exists one). We have now

Theorem 3. *It holds* $\gamma_k = (0, \ldots , 0, 1)\, E_k\, (\# A, \ldots , \# A^k)^T$ *(for* $k > 0$*).*

Thus γ_k can be expressed as a linear combination of $\# A, \ldots , \# A^k$ the coefficients of which do not depend on $\# A$.

Denote

$$D_{k,i} = \{P \in A^{(*)} \mid |P| = k, \exp(P) = i\} \qquad (\text{for } k, i > 0) .$$

Further denote by δ_k the number of $(\omega^* + \omega)$-powers of length k over A (for $k > 0$).

Theorem 4. $\delta_k = \# D_{k,1}$ *(for* $k > 0$*).*

Proof. For each circular word P, $P^{\Sigma'}$ is an $(\omega^* + \omega)$-power of length k if and only if $\sqrt{P}$ is in $D_{k,1}$. This follows by Lemma 2 and Theorem 1. Theorem 1 also implies that if $\sqrt{P} \neq \sqrt{Q}$, then $P^{\Sigma'} \neq Q^{\Sigma'}$. $\blacksquare$

To derive a formula for $\# D_{k,1}$ we apply the analogue of Lemma 4 to circular words.

Lemma 5. $\sum_{m|k} \# D_{m,1} = \gamma_k$ *(for* $k > 0$*).*

Proof. Clearly

$$\bigcup_{m|k} \{P \in A^{(*)} \mid |P| = k, |\sqrt{P}| = m\} = \{P \in A^{(*)} \mid |P| = k\}$$

and the union is disjoint. If $|P| = |Q| = k$, $|\sqrt{P}| = |\sqrt{Q}| = m$, $\sqrt{P} \neq \sqrt{Q}$ and $P = Q$, a contradiction immediately arises by Lemma 2, since $\exp(\sqrt{P}) = \exp(\sqrt{Q}) = 1$. Thus

$$\# \{P \in A^{(*)} \mid |P| = k, |\sqrt{P}| = m\} = \# D_{m,1}. \blacksquare$$

Theorem 5. $\delta_k = (0, \ldots , 0, 1)\, S_k^{-1} (\gamma_1, \ldots , \gamma_k)^T$.

Proof. Theorem 4 and Lemma 5. ▮

Denote

$$\varepsilon_i = \big((0, \ldots, 0, 1)\, E_i \mid 0, \ldots, 0\big) \qquad (\text{for } i = 1, \ldots, k)$$

(a k-vector) and

$$\bar{E}_k = \begin{pmatrix} \varepsilon_1 \\ \cdot \\ \cdot \\ \varepsilon_k \end{pmatrix}.$$

Combining Theorems 3 and 5 we get the following formula for δ_k.

Theorem 6. $\delta_k = (0, \ldots, 0, 1) S_k^{-1} \bar{E}_k (\# A, \ldots, \# A^k)^T$ (*for* $k > 0$).

3. Some generative properties

Bearing in mind the intuitive interpretation of circular words as "circular sequences" of symbols it is obvious how Chomsky type grammars should be defined so as to generate circular words starting from some initial nonterminal circular word. The families of languages of circular words thus obtained are denoted by REG_Ω, CF_Ω, CS_Ω and RE_Ω for type 3, 2, 1 and 0 grammars, respectively (cf. Salomaa [3]). These families are easily characterized. For a family F of languages of "usual" words we denote

$$F^\Omega = \{L^\Omega \mid L \text{ is in } F\}$$

where

$$L^\Omega = \{P^\Omega \mid P \text{ is in } L\}.$$

Theorem 7. (i) $\mathrm{REG}_\Omega = \mathrm{REG}^\Omega$, (ii) $\mathrm{CF}_\Omega = \mathrm{CF}^\Omega$, (iii) $\mathrm{CS}_\Omega = \mathrm{CS}^\Omega$, (iv) $\mathrm{RE}_\Omega = \mathrm{RE}^\Omega$

Proof. (i) and (ii) are clear. The proof of (iii) and (iv) (is by standard techniques (details are omitted here). ▮

In (i) and (ii) the crucial factor is context-free rewriting. In (iii) and (iv) it is context-sensitive rewriting together with the use of nonterminal "blocks" and "messengers". Thus interesting behaviour can be expected concerning grammars which use context-sensitive type rewriting but no nonterminals. Indeed, for IL systems (see Herman and Rozenberg [1]; generation of circular words by these systems is defined in the obvious way) the situation is quite different from the one in Theorem 7.

Using a notation as above we have

Theorem 8. IL_Ω *and* IL^Ω *are incomparable.*

Proof. Since $\mathrm{UIL}^\Omega = \mathrm{UOL}^\Omega$, any unary IL language L which is not in UOL (say $\{a, a^2\}$) is such that L^Ω is not in IL_Ω.

Clearly the language $L' = L_1 \cup L_2$ where

$$L_1 = \{(a^{2m} c a^{2^n - m} b a^{3^n} b)^3 \mid 0 \leqq m \leqq 2^n, n \geqq 0\},$$

$$L_2 = \{(a^{3m} d a^{3^n - m} b a^{2^{n+1}} b)^3 \mid 0 \leqq m \leqq 3^n, n \geqq 0\}$$

is such that $(L')^\Omega$ is in IL_Ω. We claim that there exists no such IL language L that $L^\Omega = (L')^\Omega$.

Assume the contrary of the above claim, i.e. $(L')^\Omega = L^\Omega$ for a language L generated by the $\langle k, h\rangle$ system $G = \langle \{a, b, c, d\}, P, g, \alpha\rangle$. Clearly any Q such that $\langle a^k, a, a^h\rangle \to_P Q$ is in $\{a\}^*$. Also clearly $Q_1 \cdots Q_{k+h+1}$ is in $\{a\}^* \{b\} \{a\}^*$ for any $Q_1, \ldots, Q_{k+h+1}$ such that

$$\langle a^k, a, a^{h-1} b\rangle \to_P Q_1, \ldots, \langle b a^{k-1}, a, a^h\rangle \to_P Q_{k+h+1}.$$

The above mentioned productions must be deterministic. Thus there exist unique numbers $i > 0$ and $j, s \geqq 0$ such that $\langle a^k, a, a^h \rangle \rightarrow_P a^i$ and $Q_1 \cdots Q_{k+h+1} = a^j b a^s$. Further, there exist unique numbers p_c, q_c and p_d, q_d such that

$$\langle a^k, a, a^{h-1}c \rangle \rightarrow_P R_1, \ldots, \langle ca^{k-1}, a, a^h \rangle \rightarrow_P R_{k+h+1} ,$$
$$\langle a^k, a, a^{h-1}d \rangle \rightarrow_P S_1, \ldots, \langle da^{k-1}, a, a^h \rangle \rightarrow_P S_{k+h+1}$$

and $R_1 \cdots R_{k+h+1} = a^{p_c} x a^{q_c}$, $S_1 \cdots S_{k+h+1} = a^{p_d} y^{q_d}$ where x, y are in $\{c, d\}$.

Let $2m', 2^{n'} - m' > k + h$. Now if $x = d$, we have

$$\begin{cases} i(2^{n'} - m' - k - h) + j + q_c = 3^n - m , \\ i(3^{n'} - k - h) + s + j = 2^{n+1} \end{cases}$$

for some n and m. For large values of n' these equations are in contradiction. So $x = c$. Similarly $y = d$.

Let $2m', 2^{n'} - m', 3u', 3^{v'} - u' > k + h$. Then we have

$$\begin{cases} i(2m' - k - h) + s + p_c = 2m , \\ i(2^{n'} - m' - k - h) + j + q_c = 2^n - m , \\ i(3^{n'} - k - h) + s + j = 3^n \end{cases}$$

and

$$\begin{cases} i(3u' - k - h) + s + p_d = 3u , \\ i(3^{v'} - u' - k - h) + j + q_d = 3^v - u , \\ i(2^{v'+1} - k - h) + s + j = 2^{v+1} . \end{cases}$$

For values of n' and v' large enough the third equations imply that $s + j = k + h$ and $i = 1$. So

$$\begin{cases} 2m' + p_c - j = 2m , \\ m' + s - q_c = m; \end{cases} \qquad \begin{cases} 3u' + p_d - j = 3u , \\ u' + s - q_d = u \end{cases}$$

and further

$$\begin{cases} p_c - j = 2(s - q_c) , \\ p_d - j = 3(s - q_d) . \end{cases}$$

In order to make L infinite we must have $p_c + q_c + s + j > 2k + 2h$ and $p_d + q_d + s + j > 2k + 2h$, i.e. $p_c > j$ and $p_d > j$.

The above results imply that whenever a symbol c or d is not in the neighbourhood of a symbol b, it moves to the right with respect to the b's in the scanned word. Since L is infinite, for each n large enough some word in

$$\bigcup_{m=2^n-k}^{2^n} \{(a^{2m}ca^{2^n-m}ba^{3^n}b)^3\}^{\Omega}$$

must directly derive a word containing occurrences of d and some word in

$$\bigcup_{m=3^n-k}^{3^n} \{(a^{3m}da^{3^n-m}ba^{2^{n+1}}b)^3\}^{\Omega}$$

must directly derive a word containing occurrences of c. This process clearly should be deterministic.

It is not difficult to see that the situation is now untenable. To maintain the above described process a synchronization of certain events at the ends of the scanned words is clearly required. However, there are no means of accomplishing this. An imaginable (but failing) procedure would be the following. Let the b's move to the left with such a speed that a symbol b reaches the left end of the word whenever a symbol c or d reaches the right end of the word. In this case the b does not "know" which symbol c or d is at the right end of the word and thus cannot "act" in the right way. ▮

As mentioned in the above proof $\mathrm{UIL}_\Omega = \mathrm{UOL}_\Omega$ and thus UIL_Ω is completely characterized (see Herman and Rozenberg [1]). It is easy to show that

$$\langle k, h\rangle_\Omega \underset{\neq}{\subset} \langle k', h'\rangle_\Omega \quad \text{iff } k + h < k' + h'$$

and that

$$\langle k, h\rangle_\Omega = \langle k', h'\rangle_\Omega \quad \text{if } k + h = k' + h'$$

whence the hierarchy of IL_Ω languages is simpler than that of IL languages. Obviously $\mathrm{OL}_\Omega = \mathrm{OL}^\Omega$.

We have given the proof of Theorem 8 in some detail as it illustrates the following peculiar feature concerning circular words and context-sensitive rewriting. Synchronizing information can go "round" the circular word in a way which cannot be simulated by back and forth motion of information in "usual" words unless nonterminals are used.

4. Circular closure and $(\omega^* + \omega)$-power closure

The *circular closure* of a language L is

$$\Omega(L) = \bigcup_{P \in L} P^\Omega$$

and the $(\omega^* + \omega)$*-power closure* of L is

$$\Sigma(L) = \bigcup_{P \in L} P^\Sigma .$$

Theorem 9. *For each regular language L, also $\Omega(L)$ is regular.*

Proof. We claim first the following. For each regular language L there exists a finite set $\varphi(L)$ of pairs of regular expressions such that, for all words P_1 and P_2, P_1P_2 is in L if and only if there exists a pair (α, β) in $\varphi(L)$ with $P_1 \in |\alpha|$ and $P_2 \in |\beta|$. The claim is proved by induction on n, the total number of applications of $+$, catenation and $*$ in the regular expression defining L (any of them can be chosen for the purpose).

Let $n = 0$. Then either L is empty or $L = \{\lambda\}$ (λ is the empty word) or $L = \{a\}$ for some symbol a. For empty L we choose $\varphi(L)$ to be empty, too. Further we take $\varphi(\{\lambda\}) = \{(\lambda, \lambda)\}$ and $\varphi(\{a\}) = \{(a, \lambda), (\lambda, a)\}$. Assume then that the claim holds true for $n < k + 1$ and let $L = |\gamma|$ where γ contains $k + 1$ applications of regular operations. Now if $\gamma = \gamma_1 + \gamma_2$ we can obviously take

$$\varphi(L) = \varphi(|\gamma_1|) \cup \varphi(|\gamma_2|) .$$

Further, if $\gamma = \gamma_1\gamma_2$ we take

$$\varphi(L) = \{(\alpha, \beta\gamma_2) \mid (\alpha, \beta) \in \varphi(|\gamma_1|)\} \cup \{(\gamma_1\alpha, \beta) \mid (\alpha, \beta) \in \varphi(|\gamma_2|)\} .$$

Finally, if $\gamma = \gamma_1^*$ we take

$$\varphi(L) = \{(\gamma_1^*\alpha, \beta\gamma_1^*) \mid (\alpha, \beta) \in \varphi(|\gamma_1|)\} \cup \{(\lambda, \lambda)\} .$$

The theorem follows from the claim since

$$\Omega(L) = \Big| \sum_{(\alpha,\beta) \in \varphi(L)} \beta\alpha \Big| . \;▮$$

Lemma 6. *Let R be a regular language and h a morphism. Then the language*

$$\bar{R} = \bigcup_{P_1P_2 \in R} \{P_2\}\,(P_1h^{-1})$$

is regular.

Proof. Let $\varphi(R)$ be given by the previous proof. Then

$$\bar{R} = \Big| \sum_{(\alpha,\beta) \in \varphi(R)} \beta\bar{\alpha} \Big| ,$$

where $|\bar{\alpha}| = |\alpha|\, h^{-1}$. ▮

Lemma 7. *Let D be a Dyck language. Denote*

$$\bar{D} = \{P_2\bar{P}_1 \mid P_1P_2 \text{ is in } D\} ,$$

where we get $\bar{P}_1$ from P_1 by replacing each parenthesis x with $\bar{x}$ or $\underline{x}$ according to whether or not its counterpart (we consider P_1 as a prefix of P_1P_2) is in P_1. Then $\bar{D}$ is a CF language.

Proof. Take the standard PDA accepting D (acceptance by empty store) and embed in it the following additional rules of behaviour. If x is a left parenthesis and y is the corresponding right parenthesis, then

$$y\bar{x} \to \lambda , \quad \underline{x}y \to \lambda .$$

This behaviour replaces the earlier one permanently when the PDA reads for the first time a symbol $\bar{x}$ or $\underline{x}$ in its input tape whereafter reading an "unbarred" symbol causes the PDA to write some nonerasable symbol on its push-down tape which then cannot be emptied. It is easily seen that the resulting PDA accepts $\bar{D}$. ▮

Theorem 10. *For each CF language L, also $\Omega(L)$ is in CF.*

Proof. Let $L = (D \cap R)\, k$ where k is a morphism, D is a Dyck language and R is a regular language. Let $\bar{D}$ and $\bar{R}$ be given by the previous lemmas (in Lemma 6 we take h to be such that $\bar{x}h = \underline{x}h = x$). The language $\bar{D} \cap \bar{R}$ is thus a CF language and so is $(\bar{D} \cap \bar{R})\, h' = \Omega(D \cap R)$ where h' is obtained from h by the additional definition $xh' = x$. Denote by $\bar{k}$ the marked morphism given by $x\bar{k} = \# xk$. Then $(\Omega(D \cap R))\, \bar{k}$ is a CF language. Starting from the PDA accepting this language it is straightforward to construct a PDA accepting $\Omega((\Omega(D \cap R))\, \bar{k})$, only a finite control is needed to take care of words not beginning with $\#$. The theorem now follows since we get $\Omega(L)$ by erasing all $\#$'s in the words of $\Omega((\Omega(D \cap R))\, \bar{k})$. ▮

Theorem 11. (i) *For each CS language L, also $\Omega(L)$ is in* CS.

(ii) *For each RE language L, also $\Omega(L)$ and $\Sigma(L)$ are in* RE.

Proof. (ii) is clear. The proof of (i) is by standard techniques (details are omitted here). ▮

As to the $(\omega^* + \omega)$-power closure of REG and CF it is easily seen that

$$\Sigma(|a^*b|) = \{(a^{n_2}ba^{n_1})^m \mid n_1, n_2 \geqq 0 \text{ and } m \geqq 1\}$$

is not in CF and thus we have.

Theorem 12. $\Sigma(\text{REG}) \not\subseteq \text{CF}$.

Denote by $\sqrt{L}$ the *root closure* of the language L, i.e.

$$\sqrt{L} = \{\sqrt{P} \mid P \text{ is in } L\}$$

(correspondingly $\sqrt{F}$ for a family F of languages). Then

$$\Sigma(L) = \mathcal{P}(\sqrt{\Omega(L)}) = \mathcal{P}(\Omega(\sqrt{L})) = \Omega(\mathcal{P}(\sqrt{L}))$$

where $\mathcal{P}$ denotes the *power closure*, i.e.

$$\mathcal{P}(L) = \{P^n \mid P \text{ is in } L \text{ and } n > 0\} .$$

It is easy to see that $\mathcal{P}(\text{CS}) \subseteq \text{CS}$. We now show that $\sqrt{\text{CS}} \nsubseteq \text{CS}$ and $\Sigma(\text{CS}) \nsubseteq \text{CS}$.

Lemma 8. *For every RE language L there exists a CS language L' and a letter-to-letter morphism h such that $L = \sqrt{L'}\, h$.*

Proof. Let $\bar{L}$ be the language which we get by replacing in each word of L its first letter x by $\bar{x}$. Then $\sqrt{\bar{L}} = \bar{L}$. It is straightforward to see that for each word P of $\bar{L}$ there exists a nonempty set I_P of positive integers such that the language

$$\{P^i \mid P \text{ is in } \bar{L} \text{ and } i \text{ is in } I_P\}$$

is a CS language (cf. Theorem 9.9 in [3]) whence the lemma follows. ▮

Corollary 2. *The family of CS languages is not closed under root closure.*

Corollary 3. *The family of CS languages is not closed under power-root closure.*

Proof. Take a CS language L such that $\sqrt{L}$ is not in CS. Assume that $\mathcal{P}(\sqrt{L}) \in \text{CS}$. Starting from the CS grammar G generating $\mathcal{P}(\sqrt{L})$ another CS grammar can be constructed which nonterminally simulates G and then after deriving a nonterminal image of a word of $\mathcal{P}(\sqrt{L})$ checks whether or not its exponent is equal to 1. If the answer is affirmative the derivation terminates with the word, if not then the derivation never terminates. Formalization of the above procedure involves a number of easy details which we will not delve into here. The latter grammar generates $\sqrt{L}$, a contradiction. ▮

Theorem 12. *The family of CS languages is not closed under $(\omega^* + \omega)$-power closure.*

Proof. Take a language L in RE $-$ CS with first symbols of words marked as in the proof of Lemma 8. Then $\Omega(L)$ is in RE $-$ CS, too. (A CS grammar generating $\Omega(L)$ could be converted into a CS grammar generating L; only those words which begin with a marked symbol can be terminal.) By Lemma 8, $\Omega(L) = \sqrt{L'}$ for some CS language L' such that $L' = \Omega(L')$ (no letter-to-letter morphism is needed here because $\sqrt{\Omega(L)} = \Omega(L)$). As in the proof of the previous corollary it is seen that

$$\Sigma(L') = \mathcal{P}(\sqrt{\Omega(L')}) = \mathcal{P}(\sqrt{L'})$$

is not in CS. ▮

It may be noted that $\sqrt{\text{REG}} \nsubseteq \text{REG}$ and $\sqrt{\text{CF}} \nsubseteq \text{CF}$. The first noninclusion follows since $\sqrt{|(a^+b)^2|}$ is not in REG (otherwise

$$|(a^+b)^2| - \sqrt{|(a^+b)^2|} = \{a^nba^nb \mid n > 0\} \in \text{REG}) .$$

The second follows since

$$\sqrt{\{ba^nb^2a^mb^2a^nb \mid n, m > 0\}}$$
$$= \{ba^nb^2a^mb^2a^nb \mid n, m > 0 \text{ and } n \neq m\} \cup \{ba^nb \mid n > 0\}$$

is not in CF. (This in turn follows by the "*uvwxy* Theorem" (see [2]) with $|uvwxy| > p$ and $|vwx| \leqq q$: Consider a word

$$P = ba^{n+q!}b^2a^nb^2a^{n+q!}b\,, \quad n > q$$

with $|P| > p$. It must be the case that its subwords u and y contain $ba^{n+q!}b$ as their common subword. The Theorem then implies that P cannot be in our language unless this is not in CF.)

Open problem

Owing to the closure of Ω and Σ under morphisms, the following question can be proposed in analogy to the question of recursive solvability of the Post correspondence problem (for which the answer is well known to be negative). Let A be a finite nonunary alphabet. Does there exist an algorithm which for any given finite alphabet B and morphisms $h_1, h_2: B^* \to A^*$ decides whether or not there exists a nonempty circular word (resp. $(\omega^* + \omega)$-power) P such that $Ph_1^\Omega = Ph_2^\Omega$ (resp. $Ph_1^\Sigma = Ph_2^\Sigma$ in analogous notation)?

References

[1] *Herman, G. T., G. Rozenberg,* Developmental Systems and Languages. North-Holland Publ. Comp., Amsterdam 1975.
[2] *Hopcroft, J. E., J. D. Ullman,* Formal Languages and Their Relation to Automata. Addison-Wesley Publ. Comp., Reading, Mass. 1969.
[3] *Salomaa, A.,* Formal Languages. Academic Press, New York 1973.

Abstract

Circular words and $(\omega^* + \omega)$-powers of words are defined as equivalence classes of words. Formulae for the number of these words of given length are proved. Circular words in connection with grammars (especially L systems) are discussed. It is shown that the union of the circular words represented by the words of a language L does not have a higher position in the Chomsky hierarchy than L and that this is not case for $(\omega^* + \omega)$-powers.

Kurzfassung

„Zirkularwörter" und „beidseitig unendliche Wörter" werden als Äquivalenzklassen von Wörtern definiert. Es werden Formeln für die Anzahl dieser „Wörter" mit gegebener Länge angegeben und Zusammenhänge zwischen Zirkularwörtern und Grammatiken (insbesondere L-Systemen) aufgezeigt. Die Vereinigung aller Zirkularwörter, deren Repräsentanten aus einer Sprache L stammen, bildet eine Sprache desselben Typs (im Sinne der Chomsky-Hierarchie) wie L; dies ist für beidseitig unendliche Wörter nicht der Fall.

Резюме

Кольцевые слова и $(\omega^* + \omega)$-степени слов определяются как классы эквивалентности слов. Доказываются отношения для числа таких слов определенной длины. Обсуждаются кольцевые слова в связи с грамматиками (в частности с Л-системами). Доказывается, что совокупность кольцевых слов, представленных с помощью слов языка L, не имеет высшего места в иерархии Хомского чем L и что это не имеет место для $(\omega^* + \omega)$-степен.

(Received: March 7, 1976)

Author's address:
K. Ruohonen
Mathematics Department
University of Turku
20500 Turku 50
Finland

Elektronische Informationsverarbeitung und Kybernetik
EIK 13 (1977) 1/2, 13–26

Real-Time Computability of $\lfloor|\alpha|x\rfloor$ and $\lfloor x^\alpha\rfloor$

By *Ivan Korec* and *Juraj Procházka*

1. Introduction

This paper deals with the complexity of the function $\lfloor|\alpha|\ x\rfloor$ (integer part of $|\alpha|\ x$) where α is an algebraic number and of the function $\lfloor x^\alpha\rfloor$ on *Minsky machines* according to the concepts introduced in [2, 3, 4].

In the following section we give an informal definition of the Minsky machine and present concepts and results dealing with the complexity of computation on Minsky machines. The concepts *"real-time computability of a function"* and *"effective admissibility for composition of a function"* [2] are very important in this paper.

In the third section the concepts of the second section are generalized for multidimensional functions.

In the further section we introduce an arithmetic in a variable number base. This is core of the solution of the problem. We show that the time for addition, subtraction, multiplication and comparing of two numbers and division of the number by a positive integer constant in a suitable number base $z(x)$ is majorized by $(z(x))^2$.

In the fifth section we show the first main result of this paper: $\lfloor|\alpha|\ x\rfloor$ is real-time computable for algebraic α. We employ the fact that if α is the root of some polynomial $f(y) = a_0 + a_1y + \cdots + a_ny^n$, then αx is the root of the polynomial $f_x(y) = (a_0x^n) + (a_1x^{n-1})\ y + \cdots + a_ny^n$. The value $\lfloor\alpha x\rfloor$ is determinated by halving of a suitable interval.

In the last section we show the second main result: $\lfloor x^\alpha\rfloor$ is real-time computable for a rational α, $\alpha \geqq 1$, and it is effectively admissible for composition if $0 \leqq \alpha \leqq 1$. The value $\lfloor x^\alpha\rfloor$ is determinated analogously as in the preceding section by halving of a suitable interval.

We use in this paper more free formulations if there is no danger of confusion so as it is standard in the papers from the theory of algorithms area.

2. Preliminaries

The concept of the computability of functions used in this paper refers to the computability on Minsky machines [2, 4].

The *Minsky machine* consists of two parts: A potentialy infinite number of *counters* $S_0, S_1, \ldots$ and a state control with finite number of states chosen from the set $\{q_0, q_1, \ldots\}$.

A pair $(q_i; m_0, m_1, \ldots, m_n)$ is said to be a *configuration*; q_i is the current state of the control and the nonnegative integeres $m_0, m_1, \ldots, m_n$ are contents of the counters $S_0, S_1, \ldots, S_n$ respectively, and for every $k > n$ the counter S_k contains a zero.

In the following the term "machine" refers to "Minsky machine". The operation on the machine has one of the following forms:

1) an addition one to the content of some counter S_j; $\langle q_iS_jPq_r\rangle$;
2) a subtraction one from the content of some counter S_j; $\langle q_iS_jMq_r\rangle$;

3) a conditional jump on zero depending on the content of the counter S_j; $\langle q_iS_jq_kq_r\rangle$.

Performing the instruction $\langle q_iS_jPq_r\rangle$ the machine moves from the configuration $(q_i; m_0, \dots, m_n)$ into $(q_r; \overline{m}_0, \dots, \overline{m}_n)$ where $\overline{m}_j = m_j + 1$ and $\overline{m}_k = m_k$ for $k \neq j$.

Performing the instruction $\langle q_iS_jMq_r\rangle$ the machine moves from the configuration $(q_i; m_0, \dots, m_n)$ into $(q_r; \overline{m}_0, \dots, \overline{m}_n)$ where $\overline{m}_j = m_j \dot{-} 1$ and $\overline{m}_k = m_k$ for $k \neq j$.

Performing the instruction $\langle q_iS_jq_kq_r\rangle$ the machine moves from the configuration $(q_i; m_0, \dots, m_n)$ into $(q_k; m_0, \dots, m_n)$ if $m_j \neq 0$ and into $(q_r; m_0, \dots, m_n)$ if $m_j = 0$.

A *machine* is a finite set of quadruples of the forms $\langle q_iS_jPq_r\rangle$, $\langle q_iS_jMq_r\rangle$, $\langle q_iS_jq_kq_r\rangle$ that does not contain two different quadruples with the same first element. The action performed by one instruction is called a *step*.

For every positive integer $n \in N$ and every machine Z the n-ary functions φ_Z^n and τ_Z^n are defined as follows. If there is a finite computation of the machine Z from the *starting configuration* $(q_1; 0, x_1, \dots, x_n)$, that ends in the *final configuration* $(q_0; y, m_1, \dots, m_u)$ for some $y, u, m_1, \dots, m_u$, then

$$\varphi_Z^n(x_1, \dots, x_n) = y\,,$$

else φ_Z^n is not defined. The function τ_Z^n is equal to the number of steps in this computation and τ_Z^n is said to be the *time complexity of the computation*.

The following theorem shows a relation between φ_Z^n and τ_Z^n.

Theorem 2.1. *Let n be a nonnegative integer and let Z be a machine. Then for every $(x_1, \dots, x_n) \in N^n$ such that $\varphi_Z^n(x_1, \dots, x_n)$ is defined,*

$$\varphi_Z^n(x_1, \dots, x_n) \leqq \tau_Z^n(x_1, \dots, x_n)$$

holds.

A function f is said to be *almost constant*, if there is $a \in N$ such that for all $(x_1, \dots, x_n) \in N^n$

$$f(x_1, \dots, x_n) = f(\min(x_1, a), \dots, \min(x_n, a))\,.$$

Theorem 2.2. *Let f be an n-ary function and $g_1, \dots, g_n$ be m-ary almost constant functions. Then composition $h = S^{n+1}(f, g_1, \dots, g_n)$ of the functions $f, g_1, \dots, g_n$,*

$$h(x_1, \dots, x_n) = f(g_1(x_1, \dots, x_n), \dots, g_n(x_1, \dots, x_n))$$

is an almost constant function.

Definition 2.3. An n-ary function f is said to be *effectively admissible for composition*, iff there exist $n, k \in N$, $n \neq 0$ and a machine Z such that $\varphi_Z^n = f$ and for every $(x_1, \dots, x_n) \in N^n$ holds:

$$\tau_Z^n(x_1, \dots, x_n) \leqq k \cdot (1 + x_1 + \dots + x_n)\,.$$

We often write "*eac*" instead of "effectively admissible for composition".

An n-tuple of nonnegative integers $(x_1, \dots, x_n)$ is denoted by X_n and similarly for the other letters (e.g. A_n, B_n). The sum $x_1 + \dots + x_n$ is often denoted by $(X_n)^+$ or X_n^+.

Theorem 2.4. *A composition of eac-functions is an eac-function.*

Theorem 2.5. *The following functions are effectively admissible for composition:*

i) *every almost constant function except the 0-ary;*

ii) I_m^n, $x + y$, $x - y$, $\lfloor x:y\rfloor$, $\lfloor\sqrt[n]{x}\rfloor$, *rest* (x, y), $\lfloor\sqrt[n]{x}\rfloor^n$, $\lfloor x:k\rfloor$, $x \dot{-} y$.

Definition 2.6. A total n-ary function is said to be *real-time computable*, iff there exist $n, k \in N$ and a machine Z such that $\varphi_Z^n = f$ and for every $X_n \in N^n$ holds:

$$\tau_Z^n(X_n) \leqq k \cdot (1 + \varphi_Z^n(X_n))\,.$$

We often write "rtc" instead of "real-time computable".

Theorem 2.7. *A composition of rtc-functions is an rtc-function.*

Theorem 2.8. *The following functions are rtc:*

$$I_m^n,\ x+y,\ x \dot{-} y,\ \lfloor x:k \rfloor,\ \lfloor \sqrt[n]{x} \rfloor^n,\ c,\ x\cdot y\ .$$

Results presented in this section were proved in [2].

3. Generalization of Preliminary Concepts for Multidimensional Functions

The concept "*m-dimensional function*" for a natural m, $m \neq 0$, is a generalization of the usual concept "*function*". If $m, n \in N$, then an m-dimensional function means a mapping of the set N^n of n-tuples of nonnegative integers into the set N^m and we write

$$\bar{f}^m : N^n \to N^m; \qquad X_n \mapsto (f_1(X_n), \dots, f_m(X_n))\ .$$

If it is evident from the context that we deal with an m-dimensional function we shall omit the adjective "m-dimensional".

For every machine Z and numbers m, n such that $m, n \in N$ and $m \neq 0$ an n-ary function $\tau_Z^{n,m}$ and an m-dimensional n-ary function $\varphi_Z^{n,m}$ are defined now.

If there exists a finite computation of Z from the *starting configuration* $(q_1; \underbrace{0, \dots, 0}_{m\text{-times}},$ $x_1, \dots, x_n)$ to the *final configuration* $(q_0; y_1, \dots, y_m, u_1, \dots, u_p)$ for suitable $p, u_1, \dots, u_p$, then

$$\varphi_Z^{n,m}(X_n) = (y_1, \dots, y_m)\ ,$$

else $\varphi_Z^{n,m}(X_n)$ will not be defined.

The value $\tau_Z^{n,m}(X_n)$ is equal to the number of the steps needed for this computation and $\tau_Z^{n,m}(X_n)$ is said to be the *time complexity of the computation of the m-dimensional function* $\bar{f}^m$ *by* Z (or shortly *time of the computation*).

The following comparison result can be proved analogously as the similar Theorem 2.1 in [2].

Theorem 3.1. *Let $m, n \in N$, $m \neq 0$ and let Z be a machine. Then for all $X_n \in N^n$ such that $\varphi_Z^{n,m}(X_n)$ is defined,*

$$(\varphi_Z^{n,m}(X_n))^+ \leqq \tau_Z^{n,m}(X_n)\ .$$

Definition 3.2. Let $m, n \in N$, $m \neq 0$; let $g(X_n)$ be a total n-ary function, then the m-dimensional function $\bar{f}^m$ is said to be *computable in the time* g, iff there exist a positive integer k and a machine Z such that

$$\varphi_Z^{n,m} = \bar{f}^m \text{ and for all } X_n \in N^n\ \tau_Z^{n,m}(X_n) \leqq k(1 + g(X_n));$$

Then we say *the time complexity $\tau_Z^{n,m}$ is majorized by g.*

Definition 3.3. The m-dimensional function

$$\bar{f}^m : N^n \to N^m; \qquad X_n \mapsto (f_1, \dots, f_m)$$

is said to be *effectively admissible for composition* (*eac*), iff there exist $m, n, k \in N$ $m \neq 0$ and a machine Z such that for every $X_n \in N^n$

$$\varphi_Z^{n,m} = \bar{f}^m \quad \text{and} \quad \tau_Z^{n,m}(X_n) \leqq k \cdot (1 + X_n^+)\ .$$

By Definition 3.2 the *eac*-function is computable in time X_n^+.

Theorem 3.4. *Let $m, n \in N$; $m \neq 0$; let functions f_i, $i = 1, \dots, m$, be eac then the m-dimensional function*

$$\bar{f}^m : N^n \to N^m; \qquad X_n \mapsto (f_1(X_n), \dots, f_m(X_n))$$

is eac.

Proof. Constructing more complex machines from the simpler machines in proofs below, the methods and results described in [2] will be used.

Assume that there are machines $Z_1, \dots, Z_m$ computing the functions f_i, $i = 1, \dots, m$, and they have the time of the computation majorized by X_n^+. We construct a machine Z from these machines. Z will compute $\bar{f}^m$ such that in the first phase it simulates actions of the machines $Z_1, \dots, Z_m$ and then it places the results into the counters $S_0^1, S_0^2, \dots, S_0^m$. The time of the computation of Z then will be majorized by X_n^+. ∎

In the following definition the concept of composition is extended for multidimensional functions.

Definition 3.5. Let $p, q, r, m, n_i \in N$, $p \neq 0$, $r \neq 0$, $n_i \neq 0$ for $i = 1, \dots, r$; let $q = n_1 + \dots + n_r$; let us have multidimensional functions:

$$\bar{g}_i^{n_i} : N^m \to N^{n_i}, \quad i = 1, \dots, r, \quad \text{and} \quad \bar{f}^p : N^q \to N^p .$$

Then we say that the p-dimensional function $\bar{e}^p$ *is formed from the functions* $\bar{f}^p, \bar{g}_1^{n_1}, \dots, \bar{g}_r^{n_r}$ *by composition* and we write $\bar{e}^p = S^{r+1}(\bar{f}^p, \bar{g}_1^{n_1}, \dots, \bar{g}_r^{n_r})$ iff for every $X_m \in N^m$

$$\bar{e}^p(X_m) = \bar{f}^p(\bar{g}_1^{n_1}(X_m), \dots, \bar{g}_r^{n_r}(X_m)) .$$

Theorem 3.6. *A composition of multidimensional eac-functions is a multi-dimensional eac-function.*

Proof. Assume that there are machines $Z_1, \dots, Z_r$ computing functions $\bar{g}_1^{n_1}, \dots, \bar{g}_r^{n_r}$ respectively and the time of each computation is majorized by X_m^+. Assume that there is a machine Z_0 computing $\bar{f}^p$ with the time of the computation majorized by X_q^+.

The machine Z computing the function $\bar{e}^p$ will be composed from the machines $Z_0, \dots, Z_r$ so that in the first phase it simulates actions of $Z_1, \dots, Z_r$ and in the second phase it simulates an action of Z_0 which has as the input data the results of the machines $Z_1, \dots, Z_r$.

The machine Z can be constructed in such a way that the time of the first phase is majorized by X_m^+ and in the second phase is majorized by $(\bar{g}_1^{n_1})^+ + \dots + (\bar{g}_r^{n_r})^+$. By Theorem 3.1 $(\bar{g}_i^{n_i})^+ \leq \tau_{Z_i}^{m, n_i}$ and thus the time of the second phase is majorized by X_m^+ as well. ∎

4. Number System with Variable Base

This section is devoted to the arithmetic in a variable number base. For this purpose we need the functions corresponding to:

i) decomposition of the number y into $z(x)$-acid representation (see Lemma 4.1);
ii) multiplication of two numbers x' and x'' and raising the number x' to the power in $z(x)$-adic representation (see the lemmas 4.3, 4.4, 4.5 and the comment after them);
iii) addition and subtraction in this representation;
iv) division of the numbers by the integer constant in this representation;
v) comparison of two numbers in this representation;
vi) enumeration of the value of a polynomial in this representation.

Define for every $m \in N$ an $(m+2)$-dimensional function $\bar{b}^{m+2}(x, y) = (x, b_0(x, y), b_1(x, y), \dots, b_m(x, y))$ where

$$b_0(x, y) = rest\ (y, x)\ ,$$
$$b_1(x, y) = rest\ (\lfloor y : x \rfloor, x)\ ,$$
$$\dots \qquad \dots$$
$$b_m(x, y) = rest\ (\underbrace{\lfloor \dots \lfloor y : x \rfloor}_{m\text{-times}} : \dots : x \rfloor, x)\ .$$

The purpose of the function $\bar{b}^{m+2}$ is to give the x-adic representation of y. (For, if $y < x^{m+1}$, then

$$y = b_0(x, y) + b_1(x, y) \cdot x + \dots + b_m(x, y) \cdot x^m\ .)$$

Lemma 4.1. *The function $\bar{b}^{m+2}$ is eac for every $m \in N$.*

Proof. By the theorems 2.4 and 2.5 the functions $I_1^1(x) = x$ and $b_i(x, y)$ for $i = 0, \dots, m$ are *eac*. Then by Theorem 3.4 $\bar{b}^{m+2}$ is *eac*. ▌

Corollary 4.2. *If $z(x)$ is eac, then the $(m+2)$-dimensional function $b^{m+2}(z(x), y)$ for every $m \in N$ is eac.*

Proof. The assertion follows immediately from the Theorems 2.4, 2.5, 3.6 and Lemma 4.1. ▌

The function for multiplication of two numbers in the $z(x)$-adic representation is constructed from two simpler functions:

i) $\bar{q}^m$ models a multiplication of two m-tuples similar to the multiplication of two polynomials (see Lemma 4.3).

ii) $\bar{r}^m$ transforms a product of $\bar{q}^m$ into the correct form in the specified number base (see Lemma 4.4).

Define for every $m \in N$, $m > 1$ the function $\bar{q}^m : N^{2m} \to N^m$; $(X_m, Y_m) \mapsto (q_1, q_2, \dots, q_m)$, where

$$q_1(X_m, Y_m) = x_1\ ,$$
$$q_2(X_m, Y_m) = x_2 y_2\ ,$$
$$q_3(X_m, Y_m) = x_2 y_3 + x_3 y_2\ ,$$
$$\dots \qquad \dots$$
$$q_m(X_m, Y_m) = x_2 y_m + x_3 y_{m-1} + \dots + x_m y_2\ .$$

Lemma 4.3. *For every $m \in N$, $m > 1$, the function $\bar{q}^m$ is computable in time $\max_{i,j} x_i x_j$.*

Proof. By the Theorems 2.7 and 2.8 the functions $q_1, \dots, q_m$ are *rtc*, hence they are computable in the time $\max_{i,j} x_i y_j$. ▌

Define for every $m \in N$, $m > 1$ the m-dimensional function $\bar{r}^m : N^m \to N^m$; $X_m \mapsto (r_1, r_2, \dots, r_m)$, where

$$r_1(X_m) = x_1\ ,$$
$$r_2(X_m) = rest\ (x_2, x_1)\ ,$$
$$r_3(X_m) = rest\ (x_3 + \lfloor x_2 : x_1 \rfloor, x_1)\ ,$$
$$\dots \qquad \dots$$
$$r_m(X_m) = rest\ (x_m + \lfloor (x_{m-1} + \lfloor (\dots + \lfloor x_2 : x_1 \rfloor) : \dots \rfloor) : x_1 \rfloor, x_1)\ .$$

Lemma 4.4. *For every $m \in N$, $m > 1$, the function $\bar{r}^m$ is eac.*

Proof. The assertion follows immediately from the theorems 2.4, 2.5 and 3.4. ∎

The composition of the functions $\bar{r}^m$ and $\bar{q}^m$ gives the m-dimensional function $\bar{h}_1^m$ that leads to the multiplication in the $z(x)$-adic representation.

Lemma 4.5. *For every $m \in N$, $m > 1$ the m-dimensional function $\bar{h}_1^m(X_m, Y_m) = \bar{r}^m(\bar{q}^m(X_m, Y_m))$ is computable in the time $\max_{i,j} x_i y_j$.*

Proof. Construct the machine Z computing $\bar{h}_1^m$ such that Z simulates $\bar{q}^m$ in the first phase and $\bar{r}^m$ in the second phase. By Lemma 4.3 is the time of the computation of the 1-st phase majorized by $\max_{i,j} x_i y_i$ and by Lemma 4.4 the time of the computation of the 2-nd phase is majorized by $(\bar{q}^m)^+$, hence by $\max_{i,j} x_i y_j$. ∎

If there is $x' \cdot x'' < (z(x))^{m-1}$ and $X'_m = \bar{b}^m(z, x')$ and $X''_m = \bar{b}^m(z, x'')$, then the value of $\bar{h}_1^m(X'_m, X''_m)$ is the product $x' \cdot x''$ in a $z(x)$-adic representation and thus

$$\bar{h}_1^m(X'_m, X''_m) = \bar{b}^m(z(x), x' \cdot x'') .$$

Clearly the time of the computation of $\bar{h}_1^m$ on a suitable Z is majorized by $(z(x))^2$. Write

$$\bar{h}_1^m(X_m, Y_m) = X_m \odot Y_m .$$

The operation $\odot$ is not commutative nor associative in N^m. It is obvious that if the first components of the m-tuples X_m, Y_m, Z_m are the same, then

$$X_m \odot Y_m = Y_m \odot X_m ,$$

but the equation

$$(X_m \odot Y_m) \odot Z_m = X_m \odot (Y_m \odot Z_m)$$

is not true in general. However, if $X'_m = \bar{b}^m(z, x')$, $X''_m = \bar{b}^m(z, x'')$, $X''' = \bar{b}^m(z, x''')$, then

$$(X'_m \odot X''_m) \odot X'''_m = X'_m \odot (X''_m \odot X'''_m) .$$

We employ the repeated application of the function $\bar{h}_1^m(X_m, Y_m)$ to obtain the power of y. Define the function $\bar{J}^m(X_m, Y_m) = X_m$ for every $m \in N$, $m \neq 0$. Then for every, $m, n \in N$, $m > 1$, $n \geqq 1$ the m-dimensional function

$$\bar{h}_n^m(X_m, Y_m) = \bar{h}_1^m(\bar{J}^m(X_m, Y_m) , \bar{h}_{n-1}^m(X_m, Y_m))$$

is computable in time $\max_{i,j} x_i y_j$. This follows immediately considering the composition of the machines in [2] and Lemma 4.5.

If now $y^n < z^{m-1}$ and $Y_m = \bar{b}^m(z, y)$, $1_m = \bar{b}^m(z, 1)$ i.e. $1_m = (z, 1, 0, \dots , 0)$ then $\bar{h}_n^m(Y_m, 1_m) = \bar{b}^m(z, y^n)$ and we denote

$$\bar{h}_n^m(Y_m, 1_m) = (Y_m)^n .$$

From above it is convenient to introduce

$$(X_m)^0 = (x_1, 1, 0, \dots , 0) \in N^m ,$$

where x_1 is the first component of X_m.

As in the Lemmas 4.3 and 4.4 we can readily obtain for $m \in N$, $m > 1$ the functions:

i) $X_m \oplus Y_m$ for the addition,
ii) $X_m \ominus Y_m$ for the recursive (arithmetical) subtraction and
iii) $\lfloor X_m \odot k \rfloor$ for the division by the positive integer constant k, such that these functions are *eac*.

The following function performs comparison of two m-tuples with the same first component (two numbers in suitable base).

Define for every $m \in N$, $m > 1$

$$Sg(X_m) = sg(x_2 + x_3 + \cdots + x_m)$$

then it is evident that the function

$$gr(X_m, Y_m) = S_g(X_m \ominus Y_m)$$

is *eac*. We write

$$X_m > Y_m\,, \quad \text{if } gr(X_m, Y_m) = 1;$$
$$X_m \leq Y_m\,, \quad \text{if } gr(X_m, Y_m) = 0\,.$$

The function modeling the enumeration of the polynomial is introduced now. The polynomial is employed to obtain the function $\lfloor \alpha x \rfloor$ in the real time, where α is a positive algebraic number. Any algebraic number α is a root of a corresponding polynomial $f(y) = a_0 + a_1 y + \cdots + a_n y^n$ with integer coefficients. Then the product αx is a root of the special polynomial $f_x(y)$ such that

$$f_x(y) = a_0 x^n + (a_1 x^{n-1})\, y + \cdots + (a_{n-1} x)\, y^{n-1} + a_n y^n\,.$$

The polynomial $f(y)$ can be written as the subtracttion of polynomials $g(y)$ and $t(y)$ with positive integer coefficients

$$f(y) = g(y) - t(y)\,.$$

Thus the enumeration can be performed with positive coefficients.

Lemma 4.6. *Let $m, n \in N$, $m \neq 0$; let $A^i_m \in N^m$, $i = 0, \ldots, m$, be m-dimensional constants. Then the m-dimensional function*

$$\bar{f}^m(X_m, Y_m) = \sum_{i=0}^{n} (A^i_m \odot (X_m)^{n-i}) \odot (Y_m)^i$$

is computable in the time $\max_{i,j} x_i y_j$.

Proof. The expression $(X_m)^0$ specifies the m-tuple $(x_1, 1, 0, \ldots, 0)$ and it is obvious that for every $Y_m \in N^m$ holds $Y_m \odot (X_m)^0 = Y_m$. We have to add, multiply and raise to the power a finite number times. The assertion follows immediately by mathematical induction from Lemma 4.5 and the comments following the lemma. ∎

Now let $f(y) = a_0 + \cdots + a_n y^n$ be the polynomial with positive integer coefficients and let $a_{\max} = \max \{a_i\}_{i=0}^{n}$. Then for every $x, y \in N$ such that $a_{\max}(n+1)\,(x^n + y^n) < (z(x))^{m-1}$ the value $\bar{f}^m(X_m, Y_m)$ is the value $f_x(y)$ in a $z(x)$-adic representation.

5. Real-Time Computability of Algebraic Numbers

In this section the real-time computability of the function $\lfloor |\alpha|\, x \rfloor$ is demonstrated for every algebraic number α.

Definition 5.1. The number α is *real-time computable* iff the function $\lfloor |\alpha|\, x \rfloor$ is real-time computable.

Evidently $\lfloor |\alpha|\, x \rfloor$ is *rtc* iff it is *eac*.

The following theorem is the main theorem of this section.

Theorem 5.2. *Every algebraic number is real-time computable.*

The theorem will be proved by constructing a machine Z computing $\lfloor|\alpha|\, x\rfloor$ in the time $\lfloor|\alpha|\, x\rfloor$. The machine Z will be composed from some simpler machines.

For every algebraic number α there exists a polynomial $f(y)$ with integer coefficients such that $f(\alpha) = 0$ and the derivative $f'(\alpha) > 0$. Hereafter the polynomial $f(y)$ will have these properties.

Without loss of generality we consider only positive algebraic numbers α and thus we consider the function $\lfloor\alpha x\rfloor$ below. Following considerations can be easy generalized for every algebraic number.

Lemma 5.3. *Let α be an algebraic number and $f(y)$ the relevant polynomial. Let $g(y)$, $t(y)$ be polynomials with positive integer coefficients such that $f(y) = g(y) - t(y)$.*

Then there exist $p, q, r, s \in \mathbb{N}$, $q \neq 0$, $s \neq 0$ such that for every $x \geqq \max\{q, s\}$ αx is the only root of the polynomial $f_x(y)$ in the interval

$$\langle \lfloor (p/q)\, x\rfloor,\ \lfloor (r/s)\, x\rfloor + 1\rangle .$$

Proof. Let $p, q, r, s \in \mathbb{N}$ be such that if $\alpha \in \langle p/q, r/s\rangle$ then α is the only root of $f(y)$ with the property $\alpha \in \langle p - 1/q, (r+1)/s\rangle$. It is evident that such p, q, r, s exist. We prove the assertion by contradiction.

Let $\beta \neq \alpha$ and $f_x(\beta x) = 0$ (see the comment preceding Lemma 4.6) and let $\beta x \in \langle \lfloor (p/q)\, x\rfloor, \lfloor (r/s)\, x\rfloor + 1\rangle$. From this $\beta x \in \langle (p/q)\, x - 1, (r/s)\, x + 1\rangle$. From the condition $x \geqq \max\{q, s\}$ it follows that $\beta \in \langle (p-1)/q, (r+1)/s\rangle$ and this is a contradiction. ∎

Corollary 5.4. *The number $\lfloor\alpha x\rfloor$ belongs to the interval*

$$\langle \lfloor (p/q)\, x\rfloor,\ \lfloor (r/s)\, x\rfloor + 1\rangle .$$

Proof. The proof is evident from $(p/q)\, x < \alpha x < (r/s)\, x + 1$. ∎

Lemma 5.5. *It holds*

$$g_x(\lfloor\alpha x\rfloor) \leqq t_x(\lfloor\alpha x\rfloor) .$$

Proof. The assertion follows immediately from $\lfloor\alpha x\rfloor \leqq \alpha x$ by considering $f'(\alpha) > 0$. ∎

Lemma 5.3 gives the convenient bounds for the value $\lfloor\alpha x\rfloor$ and Lemma 5.5 is used as a criterion to find $\lfloor\alpha x\rfloor$. The value $\lfloor\alpha x\rfloor$ is found by dividing the interval $\langle \lfloor (p/q)\, x\rfloor, \lfloor (r/s)\, x\rfloor + 1\rangle$ into halves and by comparing "the centre" in accordance with Lemma 5.5.

Lemma 5.6. *Let $a, b \in \mathbb{N}$, $a < b$; let φ be a nondecreasing function on $\mathbb{N}$ such that $\varphi(a) = 0$ and $\varphi(b) > 0$.*

Then it is sufficient to compute $\lfloor\log_2 (b - a)\rfloor + 2$ values of the function φ to find the greatest w such that $\varphi(w) = 0$.

Proof. Let $\langle x_0, y_0\rangle = \langle a, b\rangle$, there obviously exists $k \in \mathbb{N}$ such that $2^{k-1} \leqq b - a < 2^k$.

If $y_i - x_i > 1$, then set

$$z_{i+1} = \left\lfloor \frac{x_i + y_i}{2} \right\rfloor$$

and

$$\langle x_{i+1}, y_{i+1}\rangle = \begin{cases} \langle x_i, z_{i+1}\rangle & \text{if } \varphi(z_{i+1}) > 0 ,\\ \langle z_{i+1}, y_i\rangle & \text{if } \varphi(z_{i+1}) = 0 . \end{cases}$$

If $y_i - x_i = 1$, then set

$$w = \begin{cases} x_i & \text{if } \varphi(y_i) > 0 , \\ y_i & \text{if } \varphi(y_i) = 0 . \end{cases}$$

Note that $y_0 - x_0 < 2^k$ implies $y_{k-1} - x_{k-1} \leqq 2$. It follows that holds $y_s - x_s = 1$ for $s = k - 1$ or $s = k$. ∎

We now halve the interval $\langle \lfloor (p/q)\, x \rfloor, \lfloor (r/s)\, x \rfloor + 1 \rangle$. If we set

$$\varphi_x(y) = \begin{cases} 1 & \text{if } g_x(y) > t_x(y) , \\ 0 & \text{else} , \end{cases}$$

then the greatest w such that $\varphi_x(w) = 0$ is $\lfloor \alpha x \rfloor$. The number of values $\varphi_x(y)$ that is sufficient for a determination of $\lfloor \alpha x \rfloor$ is not greater then $\lfloor \log_2 (\lfloor (r/s)\, x \rfloor + 1 - \lfloor (p/q)\, x \rfloor) \rfloor + 2$ by Lemma 5.6. It is evident that there is $k \in N$ such that

$$\left\lfloor \log_2 \left(\left\lfloor \frac{r}{s} x \right\rfloor + 1 - \left\lfloor \frac{p}{q} x \right\rfloor \right) \right\rfloor + 2 \leqq k \cdot (\lfloor \sqrt[3]{x} \rfloor + 1) .$$

By the last section the polynomial in the $z(x)$-adic representation is computable in the time $(z(x))^2$. We want to compute $\lfloor \alpha x \rfloor$ in the real time and this is equivalent to showing that $\lfloor \alpha x \rfloor$ is computable in time x. We see that the time of the computation of $\lfloor \alpha x \rfloor$ is majorized by the time of the computation of the polynomial multiplied by the number of the divisions of the interval. If we set $z(x) = \lfloor \sqrt[3]{x} \rfloor + 1$ (one is added to avoid the base $z(x) = 1$), we shall obtain that the time of the computation of $\lfloor \alpha x \rfloor$ will be majorized by $(\lfloor \sqrt[3]{x} \rfloor + 1)^3$, i.e. it is majorized by x.

Estimate now the number m of the counters needed for m-tuples of the machine Z computing $\lfloor \alpha x \rfloor$.

Let $f(y) = a_0 + a_1 y + \cdots + a_n y^n$ be the relevant polynomial with the number α and let p, q, r, s be the numbers from Lemma 5.3. If $a_{\max} = \max \{ |a_i| \}_{i=0}^{n}$, $x \geqq \max \{q, s\}$ and $k = \max \{r + 1, p\}$ then for every $y \in \langle \lfloor (p/q)\, x \rfloor, \lfloor (r/s)\, x \rfloor + 1 \rangle$ we obtain

$$f_x(y) = a_0 x^n + (a_1 x^{n-1})\, y + (a_2 x^{n-2})\, y^2 + \cdots + a_n y^n \leqq$$
$$\leqq a_{\max}(x^n + k \cdot x^n + \cdots + k^n \cdot x^n) \leqq a_{\max} \cdot k^{n+1} \cdot x^n .$$

Since polynomials $g(y)$, $t(y)$ with the property $f(y) = g(y) - t(y)$ can be selected such that the coefficients of $g(y)$ are the positive coefficients of $f(y)$ and the coefficients of $t(y)$ are the absolute value of the negative coefficients of $f(y)$, we can perform this estimation.

It is sufficient to select such m that

$$a_{\max} \cdot k^{n+1} \cdot x^n < (\lfloor \sqrt[3]{x} \rfloor + 1)^{m-1} .$$

Evidently there exists $b \in N$ such that

$$a_{\max} k^{n+1} < 2^b \leqq (\lfloor \sqrt[3]{x} \rfloor + 1)^b .$$

Thus

$$a_{\max} k^{n+1} x^n < (\lfloor \sqrt[3]{x} \rfloor + 1)^{3n+b}$$

and hence we set

$$m = 3n + b + 1 .$$

We now introduce the flowchart 1 of Z computing the function $\lfloor \alpha x \rfloor$ in the time x. Let $1_m = \bar{b}^m(z(x), 1)$ and $S_z, S_{y_0}, S_{y_1}, \ldots, S_{y_{m-2}}$ be the counters of the resulting m-tuple $(z, y_0, y_1, \ldots, y_{m-2})$.

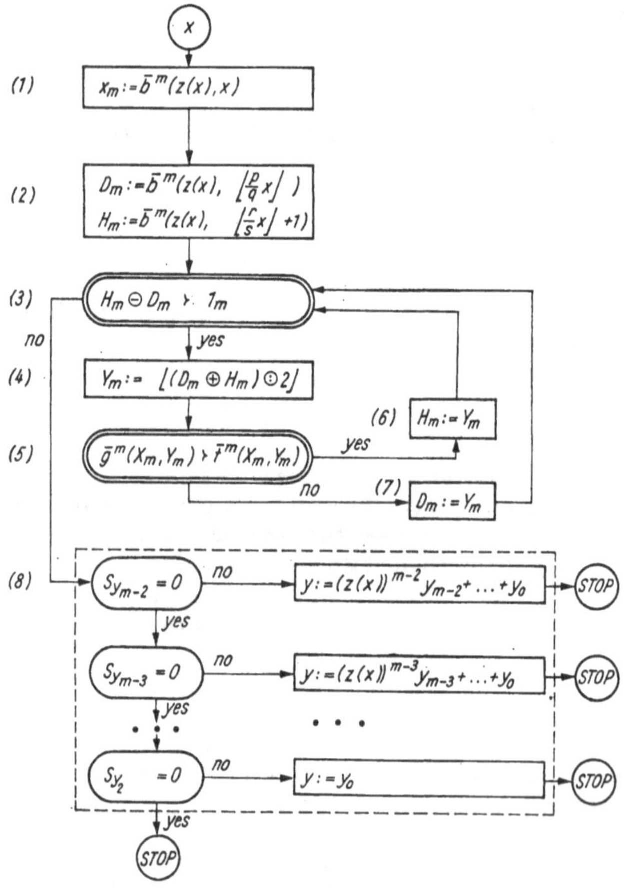

Flowchart 1

Decision blocks are distinguished by the frame. A simple frame indicates a simple decision block and a double frame the connection of the operation block with the simple decision block. Hence the blocks (3) and (5) of the Flowchart 1 are the connection of the operation block with the function $gr(X_m, Y_m)$ and of the simple decision block.

Proof of Theorem 5.2. Let α be a positive algebraic number and $p, q, r, s \in N$ where $q \neq 0$, $s \neq 0$ be the numbers from Lemma 5.3; and let $x \geqq \max\{q, s\}$. The *rtc*-property does not depend on the finite number of arguments $x \in N$. Let m be the sufficient number of counters for m-tuples.

We construct the machine Z computing $\lfloor \alpha x \rfloor$ by composing the machines corresponding to the blocks from (1) to (8) of the flowchart 1.

To pass through an operation block we must compute all functions marked in the block. To pass through a decision block we must compute all functions marked in the block and add one step for decision.

The block is said to be computable in the time $f(x)$, iff every function contained in it is computable in the time $f(x)$.

i) A system of blocks from (1) to (7) is computable in the time x, because blocks are computable in time $(\lfloor \sqrt[3]{x} \rfloor + 1)^2$, except (1) and (2) which are computable in the time x, and the blocks from (3) to (7) can be repeated at most $k \cdot (\lfloor \sqrt[3]{x} \rfloor + 1)$-times for suitable $k \in \mathbb{N}$.

ii) The system (8) is computable in time $\lfloor \alpha x \rfloor$, because the $(r + 2)$-ary function

$$f(x, y_0, \dots, y_r) = x^r y_r + x^{r-1} y_{r-1} + \cdots + x \cdot y_1 + y_0$$

is *rtc* by the theorems 2.7, 2.8. It is evident that

$$(z(x))^r y_r + (z(x))^{r-1} y_{r-1} + \cdots + y_0 = \lfloor \alpha x \rfloor .$$

Considering i) and ii) we may conclude that $\lfloor \alpha x \rfloor$ is computable in the time $\lfloor \alpha x \rfloor$. That completes the proof of Theorem 5.2.

6. Real-Time Computability of the Power

Nonnegative numbers α such that $\lfloor x^\alpha \rfloor$ is *rtc* are investigated in this section. Clearly negative α are not investigated for $\lfloor x^\alpha \rfloor$.

Let $\alpha \in (0, 1)$, then obviously for every $x > 1$, $x^\alpha < x$. Hence $\lfloor x^\alpha \rfloor$ cannot be *rtc*, because the time of the computation of Z computing $\lfloor x^\alpha \rfloor$ is not smaller then the input datum x.

The function $\lfloor x^0 \rfloor$ is a constant thus it is *rtc*.

Let now $\alpha \geqq 1$.

If α is a positive integer the function $\lfloor x^\alpha \rfloor$ will be *rtc* by Theorem 2.8 and considering $\lfloor x^\alpha \rfloor = x^\alpha$.

Let α be a rational number. It is sufficient to consider α from the open interval $(1, 2)$ because if $\lfloor x^\alpha \rfloor$ is *rtc* then the function $\lfloor x^{\alpha n} \rfloor = \lfloor (x^n)^\alpha \rfloor$ is *rtc* by the Theorems 2.7 and 2.8.

The main result of this section is the following

Theorem 6.1. *For every rational* $\alpha \geqq 1$, $\lfloor x^\alpha \rfloor$ *is rtc.*

The theorem will be proved analogously as in Section 5 by constructing from simpler machines the machine Z computing $\lfloor x^\alpha \rfloor$.

Lemma 6.2. *Let p, q be in $\mathbb{N}$ such that $\alpha = p/q$; then $y = \lfloor x^{p/q} \rfloor$ iff*

$$y^q \leqq x^p < (y + 1)^q . \tag{1}$$

Proof. Assume $y = \lfloor x^{p/q} \rfloor$, then the inequality

$$\lfloor x^{p/q} \rfloor^q \leqq x^p < (\lfloor x^{p/q} \rfloor + 1)^q$$

follows from the inequality

$$\lfloor x^{p/q} \rfloor \leqq x^{p/q} < (\lfloor x^{p/q} \rfloor + 1) .$$

If $y \neq \lfloor x^{p/q} \rfloor$ then there exists $x \in \mathbb{N}$ such that inequality (1) does not hold. ∎

We can bound $\lfloor x^{p/q} \rfloor$ for every $q > 0$ by x^p. The value $\lfloor x^{p/q} \rfloor$ is found by halving the interval $\langle 0, x^p \rangle$ and by comparing "the centre" in accordance with (1).

If we set

$$\varphi_x(y) = \begin{cases} 1 & \text{for } x^p < y^q \\ 0 & \text{else;} \end{cases}$$

then the greatest w such that $\varphi_x(w) = 0$ is $\lfloor x^{p/q} \rfloor$. By Lemma 5.6 the number of values that is sufficient to determine $\lfloor x^{p/q} \rfloor$ is not greater than $\lfloor \log_2 x^p \rfloor + 2$. Evidently there is $k \in \mathbb{N}$ such that

$$\lfloor \log_2 x^p \rfloor + 2 \leqq k(\lfloor \sqrt[3]{x} \rfloor + 1) .$$

Analogously as in Section 5 we choose $z(x) = \lfloor \sqrt[3]{x} \rfloor + 1$.

Estimate now the number m of the counters needed for m-tuples of the machine Z computing $\lfloor x^{p/q} \rfloor$ for $p/q > 0$, where $p, q \in N$. The upper bound of $\lfloor x^{p/q} \rfloor$ is x^p, hence Z operates with no number greater than x^{pq} by Lemma 6.2 and comment following it. Since $z(x) = \lfloor \sqrt[3]{x} \rfloor + 1$ then evidently $x < (z(x))^3$ and this implies

$$x^{pq} < (z(x))^{3pq} .$$

It is sufficient to set $m = 3pq + 1$.

The machine Z computing $\lfloor x^{p/q} \rfloor$ in real time is represented by the Flowchart 2. The flowchart 2 is similar to Flowchart 1 and we do not repeat comments for it.

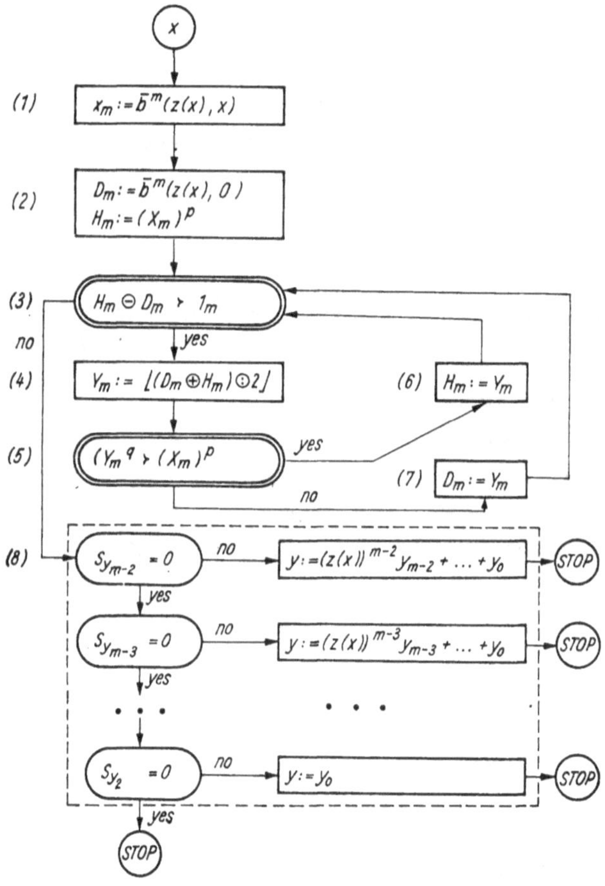

Flowchart 2

Proof of Theorem 6.1. Without loss of generality let $p/q \neq 1$ thus $p/q > 1$. The machine Z is constructed from the machines corresponding to the blocks (1) to (8) of the Flowchart 2.

i) The system of blocks (1) to (7) is computable in the time x.

ii) The system (8) is computable in the time $\lfloor x^{p/q} \rfloor$.

Thus the function $\lfloor x^{p/q} \rfloor$ is *rtc* because $\lfloor x^{p/q} \rfloor \geq x$.

Theorem 6.3. *Let α be a rational number from the interval $\langle 0, 1 \rangle$ then $\lfloor x^\alpha \rfloor$ is eac.*

Irrational numbers α such that $\lfloor x^\alpha \rfloor$ is *rtc* are not known to the authors. On the other hand, there is a recursive real number α such that $\lfloor x^\alpha \rfloor$ is not real-time computable. Indeed, consider a 0–1 valued recursive but not primitive recursive function $f(x)$, $f(0) = 1$. Then the number α which has subsequent binary development

$$f(0),\ f(1)\, f(2)\, f(3) \ldots$$

is recursive. If the function $\lfloor x^\alpha \rfloor$ is real-time computable, then it is primitive recursive by [1]. Hence

$$f(x) = \lfloor \log_2 \lfloor (2^{2^{x+1}})^\alpha \rfloor \rfloor \dot{-} 2 \cdot \lfloor \log_2 \lfloor (2^{2^x})^\alpha \rfloor \rfloor$$

is also primitive recursive and this is a contradiction. It can be shown analogously that there are primitive recursive real numbers α ($\alpha > 1$) such that $\lfloor x^\alpha \rfloor$ is not real-time computable.

Acknowledgement. At the end the authors would like to express their thanks to RNDr *J. Gruska*, CSc for his valuable advices and remarks and to PhD *B. Rovan* for his gracious reading of this paper and his valuable remarks.

References

[1] *Benešova, M., I. Korec*, Real-time Computable Functions and Almost Primitive Recursive Functions. Acta fak. rer. nat. Univ. Comenianae (Bratislava) **23** (1969).

[2] *Korec, I.*, Complexity Valuation of the Partial Recursive Functions Following the Expectation of the Length of Their Computation on Minsky Machines. Acta fak. rer. nat. Univ. Comenianae (Bratislava) **23** (1969).

[3] *Korec, J.*, Úvod do teórie algoritmov. Skriptum UK Bratislava, 1974.

[4] *Мальцев А. И.*, Алгоритмы и рекурсивные функции. Москва 1965.

Abstract

The function f is real-time computable (effectively admissible for composition) if the length of its computation on a Minsky machine is majorized by the same function f (by the sum of its arguments respectively). This paper studies the functions $\lfloor |\alpha|\, x \rfloor$ and $\lfloor x^\alpha \rfloor$ and the main results show:

i) if α is an algebraic number then $\lfloor |\alpha|\, x \rfloor$ is real-time computable.

ii) if α is a rational number $\alpha \geqq 1$ then $\lfloor x^\alpha \rfloor$ is real-time computable and if $0 \leqq \alpha \leqq 1$ then $\lfloor x^\alpha \rfloor$ is effectively admissible for composition.

Kurzfassung

Eine Funktion $f(x)$ heißt in Realzeit berechenbar, wenn es eine Minsky-Maschine gibt, die sie in der Zeit $k \cdot (1 + f(x))$ berechnet. Es wird folgendes bewiesen: In Realzeit berechenbar sind die Funktionen $\lfloor |\alpha|\, x \rfloor$ (für algebraische Zahlen α) und $\lfloor x^\alpha \rfloor$ (für rationale Zahlen $\alpha \geqq 1$).

Резюме

Функция f называется вычислимой в реальное время (эффективно допустима для суперпозиции) если длина (время) ее вычисления на машине Минского мажорируется самой функцией f (суммой ее переменных). В статье исследуются функции $\lfloor |\alpha|\, x \rfloor$ и $\lfloor x^\alpha \rfloor$ и доказываются следующие основные результаты:

1) если α алгебраическое число, тогда $\lfloor |\alpha| x \rfloor$ вычислима в реальное время.

2) если α рациональное число и $\alpha \geqq 1$ тогда $\lfloor x^\alpha \rfloor$ вычислима в реальное время, а если $0 \leqq \alpha \leqq 1$, тогда $\lfloor x^\alpha \rfloor$ эффективно допустима для суперпозиции.

(Received October 15, 1975)

Authors' addresses:

RNDr. I. Korec
Department of Algebra and Theory
of Numbers
Faculty of Science of Comenius University
Matematický pavilón
81631 Bratislava
ČSSR

RNDr. J. Procházka
Department of Computer Science
Faculty of Science of Comenius University
Matematický pavilón
81631 Bratislava
ČSSR

Elektronische Informationsverarbeitung und Kybernetik
EIK 13 (1977) 1/2, 27–38

Zur Teilwortkomplexität für Wörter und Folgen über einem endlichen Alphabet

Von *Michael Heinz*[1])

Von A. N. Kolmogoroff ist in den Arbeiten [2, 3] der Begriff der Programmkomplexität eines Wortes p eingeführt worden. Hierbei wird jedem Wort p der für dieses Wort unbedingt nötige Aufwand zur Erzeugung bzw. zur Konstruktion des Wortes als seine Komplexität zugeordnet, wobei der Begriff „Aufwand" algorithmentheoretisch präzisiert wird. Dieses Komplexitätsmaß wird erst für Wörter großer Länge hinreichend genau und ist daher für eine Reihe von praktischen Fragestellungen, bei denen Komplexitäten (z.B. Lernzeiten in der organismischen Informationsverarbeitung) von Wörtern kleinerer Länge betrachtet werden, schlecht anwendbar. In den letzten Jahren sind verstärkt Anstrengungen unternommen worden, Komplexitätsmaße festzulegen, die diese Schwäche nicht besitzen. Die Arbeit [8] von A. Lempel und J. Zif, auf die in einer späteren Arbeit näher eingegangen werden soll, ist eines der neueren Beispiele dafür.

In [4] und [5] werden von H. Thiele Komplexitätsmaße vorgeschlagen, die für praktische Problemstellungen geeignete Resultate liefern könnten. In dieser Arbeit soll ein Spezialfall eines in [5] entwickelten gruppentheoretischen Ansatzes zur Bestimmung der Komplexität der Struktur eines Wortes, seine Teilwortkomplexität, genauer untersucht werden. Psychologische Untersuchungen haben bereits Zusammenhänge zwischen der Teilwortkomplexität eines Wortes und seiner experimentell ermittelten Lernzeit festgestellt (siehe [6]).

Von A. Ehrenfeucht, K. P. Lee und G. Rozenberg wird in [1] die Teilwortkomplexität benutzt, um Charakterisierungen von determinierten Lindenmayer-Systemen ohne Kontextabhängigkeiten und speziellen Teilklassen von ihnen anzugeben. Wir kommen im letzten Abschnitt noch einmal auf diese Arbeit zurück. Sie beweist, daß dieses Komplexitätsmaß auch für die Sprachtheorie Bedeutung besitzt. Die vorliegende Arbeit soll diese These erhärten.

Im ersten Abschnitt werden die benötigten Begriffe eingeführt und einige Hilfsmittel für die weiteren Abschnitte geschaffen. Der zweite Abschnitt beinhaltet Charakterisierungssätze von Anfangsstück- bzw. Mittelstückmengen von Folgen mit Hilfe der Teilwortkomplexität. Im dritten Abschnitt wird nachgewiesen, daß die Eigenschaft eines Wortes, eine Komplexität pro Symbol kleiner oder gleich einer reellen Zahl c zu besitzen, regulär ist. Der letzte Abschnitt ist einigen Eigenschaften, die die Teilwortkomplexität bzgl. Homomorphismen besitzt, gewidmet.

Der Autor möchte an dieser Stelle Herrn Prof. Dr. H. Thiele (Berlin) für das der Arbeit entgegengebrachte Interesse und seine vielfältigen Anregungen danken.

---

[1]) Sektion Mathematik der Humboldt-Universität zu Berlin.

1. Definitionen und Eigenschaften

Wir wollen mit X^* die freie Worthalbgruppe über dem Alphabet X, mit X^n die Menge aller Wörter der Länge n aus X^*, mit e das leere Wort, mit $\mathbb{N}$ bzw. $\mathbb{N}^+$ die Menge der natürlichen Zahlen mit bzw. ohne Null, mit $\mathbb{G}$ die Menge der ganzen Zahlen, mit $X^{\mathbb{N}^+}$ bzw. $X^{\mathbb{G}}$ die Menge der Abbildungen der natürlichen Zahlen bzw. der ganzen Zahlen in die Menge X, mit $X^{[n]}$ die Menge aller Wörter der Länge kleiner oder gleich n, mit $l(p)$ die Länge eines Wortes p, mit $|M|$ die Kardinalzahl der Menge M und mit $[c]$ die größte ganze Zahl, die kleiner oder gleich der reellen Zahl c ist, bezeichnen. Wenn $\xi \in X^{\mathbb{N}^+}$, so soll $\xi[n]$ das Anfangsstück der Folge ξ der Länge n $\big(\xi[n] =_{\mathrm{Df}} = \xi(1) \cdots \xi(n)\big)$ und wenn $\eta \in X^{\mathbb{G}}$, so soll $\eta[-n, n]$ das Mittelstück der Folge η der Länge $2n + 1$ $\big(\eta[-n, n] =_{\mathrm{Df}} \eta(-n) \cdots \eta(0) \cdots \eta(n)\big)$ bezeichnen.

Definition.

a) Seien $p, q \in X^*$, dann ist q in $p =_{\mathrm{Df}} \exists p_1 \exists p_2 (p_1, p_2 \in X^* \wedge p = p_1 q p_2)$.

b) Sei $p \in X^*$ und $i \in \mathbb{N}^+$, dann ist

$$K(p, i) =_{\mathrm{Df}} |\{q \mid q \in X^* \wedge q \text{ in } p \wedge l(q) = i\}| .$$

c) Für $p \in X^*$ nennen wir

$$K(p) =_{\mathrm{Df}} \sum_{i=1}^{l(p)} K(p, i)$$

die *Teilwortkomplexität* des Wortes p.

$K(p, i)$ ist somit die Anzahl der verschiedenen Teilwörter von p der Länge i und $K(p)$ ist die Anzahl der verschiedenen Teilwörter von p, die ungleich e sind.

Die Definition von $K(\,.\,, i)$ läßt sich auf Folgen aus $X^{\mathbb{N}^+}$ bzw. $X^{\mathbb{G}}$ ausdehnen, indem wir die in-Beziehung in natürlicher Weise verallgemeinern. Wenn wir die Komplexität $K(\,.\,)$ auch auf Folgen aus $X^{\mathbb{N}^+}$ und $X^{\mathbb{G}}$ ausdehnen wollen, so stellen wir fest, daß diese Funktion dazu nicht geeignet ist, weil jede Folge als Komplexität $\aleph_0$ haben würde.

Es bietet sich hierfür die Funktion

$$N(p) =_{\mathrm{Df}} \begin{cases} \dfrac{K(p)}{l(p)}, & \text{falls } p \neq e, \\ 0, & \text{falls } p = e \end{cases}$$

eher an. Diese Größe wollen wir *Komplexität pro Symbol* eines Wortes p nennen.

Mit Hilfe der Funktion N definieren wir nun die Komplexität von Folgen:

a) $$N(\xi) =_{\mathrm{Df}} \sup_n \{N(\xi[n])\} \quad \text{für } \xi \in X^{\mathbb{N}^+},$$

b) $$N(\eta) =_{\mathrm{Df}} \sup_n \{N(\eta[-n, n])\} \quad \text{für } \eta \in X^{\mathbb{G}}.$$

Bemerkung 1. N erfüllt die folgende Ungleichung

$$1 \leqq N(p) \leqq \frac{l(p) + 1}{2} \quad \text{für alle } p \in X^* .$$

Dies ist leicht einzusehen, wenn man berücksichtigt, wieviele Teilwörter ein Wort p maximal haben kann.

Lemma 1. *Für alle $p, q \in X^*$ gelten folgende Beziehungen:*

a) $$N(pq) \leqq N(p) + l(q),$$

b) $$N(pq) \leqq N(q) + l(p).$$

Der Beweis erfolgt durch Induktion über q bzw. p und sei dem Leser überlassen.

Bemerkung 2. Es gibt keine natürliche Zahl c, so daß $N(pq) \leqq N(p) + N(q) + c$ für alle $p, q \in X^*$ gilt. Wir geben dafür folgendes Beispiel an: Sei $p = a^n$ und $q = b^n$, dann ergibt sich $N(a^n) = N(b^n) = 1$, hingegen ist $N(a^n b^n) = n/2 + 1$.

Lemma 2 (Eigenschaften der Funktion $K(.\,,.)$).

a) *Für jedes $i \in \mathbb{N}^+$ und jedes $p \in X^*$ gilt*

$$K(p, i) \leqq l(p) - i + 1 .$$

b) *Für alle $i \in \mathbb{N}^+$ und alle $p \in X^*$ gilt: Wenn $l(p) > K(p, i) + i - 1$, dann ist $K(p, i) \leqq K(p, i + 1)$.*

c) *Sei $K_{\max}(p) = \max \{K(p, i) \mid i \in \{1, \ldots, l(p)\}\}$ und i_0 die größte Zahl mit $K(p, i_0) = K_{\max}(p)$, dann gilt*

$$K(p, i_0 + j) = K_{\max}(p) - j \quad \text{für } j \in \{1, \ldots, K(p, i_0) - 1\} .$$

d) *Sei $K(p, i) = K(p, i + 1)$ und $l(p) > K(p, i) + 1$, dann ist auch $K(p, i) = K(p, i + 2)$.*

Beweis. Sei $w = l(p) - i + 1$, und sei $q_1, q_2, \ldots, q_w$ die Folge der Teilwörter und $p_1, p_2, \ldots, p_{K(p,i)}$ die Folge der verschiedenen Teilwörter der Länge i von p jeweils in der Reihenfolge ihres Auftretens (von links nach rechts). Außerdem sei ${}^{\backprime}q_i = b_i q_i$ und $q_i' = q_i a_i$, wobei b_i der Buchstabe ist, der (in p) q_i vorangeht und a_i derjenige ist, der (in p) q_i nachfolgt.

a) Die maximale Anzahl der Teilwörter der Länge i in p ist w.

b) Fall 1. Es gibt ein a, so daß $p_{K(p,i)}a$ in p ist. — Dann sind die $p_i a_i$ für $i \in \{1, \ldots, K(p, i)\}$, wobei a_i der Buchstabe ist, der dem ersten Auftreten von p_i in p folgt, verschiedene Teilwörter der Länge $i + 1$ von p.

Fall 2. Es gibt kein a mit $p_{K(p,i)}a$ ist in p. — Dann ist $p_{K(p,i)}$ Endstück von p und tritt nur einmal in p auf. Aus der Ungleichung $l(p) > K(p, i) + i - 1$ folgt somit, daß es ein j mit $1 \leqq j \leqq w$ geben muß, so daß q_j verschieden ist von allen q_i für $1 \leqq i < j$, und für q_{j-1} gilt dies nicht. Sei j_0 ein solches j. Dann sind die $p_i a_i$ (wie im Fall 1) für $i \in \{1, \ldots, K(p, i) - 1\}$ und ${}^{\backprime}q_{j_0}$ voneinander verschieden. Damit ist b) nachgewiesen.

c) Folgt leicht aus a) und b).

d) Fall 1. Sei $p_j = q_j$ für $j \in \{1, \ldots, K(p, i)\}$. — Nach $K(p, i) = K(p, i + 1)$ sind dann die Buchstaben a_i, die in p auf p_i folgen können, für $i \in \{1, \ldots, K(p, i)\}$ eindeutig bestimmt. Dies zieht aber $K(p, i) = K(p, i + 2)$ nach sich.

Fall 2. $q_1, \ldots, q_{K(p,i)}$ sind nicht voneinander verschieden. — Mit Hilfe des Beweises von b) kann man aus $K(p, i) = K(p, i + 1)$ schließen, daß die Folge der Teilwörter der Länge i von p folgende Eigenschaft besitzen muß: Es gibt ein k und ein l mit $1 \leqq k < l < w$ und

(1) $q_1, \ldots, q_k, q_{l+1}, \ldots, q_w$ sind voneinander verschieden,

(2) $q_s \in \{q_1, \ldots, q_{s-1}\}$ für $s \in \{k + 1, \ldots, l\}$.

$|\{q_1, \ldots, q_k, q_{l+1}, \ldots, q_w\}| = K(p, i)$. Die $K(p, i)$ verschiedenen Teilwörter der Länge $i + 1$ von p sind somit $q_1', \ldots, q_k', {}^{\backprime}q_{l+1}, q_{l+1}', \ldots, q_{w-1}'$. Dann sind die Wörter $q_1'', \ldots, q_k'', {}^{\backprime\backprime}q_{l+1}, {}^{\backprime}q_{l+1}', q_{l+1}'', \ldots, q_{w-2}''$ verschiedene Teilwörter der Länge $i + 2$ von p. Für ein q_t'' mit $t \in \{k + 1, \ldots, l - 2\}$ gibt es ein $m \in \{1, \ldots, k\}$, so daß $q_t' = q_m'$ erfüllt ist. Damit müssen aber auch die Endstücke der Länge i von q_t' und q_m' übereinstimmen. Wegen $K(p, i) = K(p, i + 1)$ ist die Fortsetzung von q_s für $s \in \{1, \ldots, l - 1\}$ eindeutig bestimmt, woraus $q_t'' = q_m''$ folgt. Außer den Wörtern $q_1'', \ldots, q_k'', {}^{\backprime\backprime}q_{l+1}$,

$q'_{i+1}, q''_{i+1}, \ldots, q''_{w-2}$ gibt es keine Wörter der Länge $i + 2$ in p und die Behauptung d) ist bewiesen.

Satz 1. *Für alle $p, q \in X^*$ mit p in q gilt $N(p) \leqq N(q)$.*

Beweis. Es genügt nachzuweisen, daß für jedes Wort $p \in X^*$ und alle $x \in X$ folgende Beziehungen gelten:

a) $N(px) \geqq N(p)$,

b) $N(xp) \geqq N(p)$.

a) Sei $Z = K(px) - K(p)$. Wenn $Z \geqq \frac{K(p)}{l(p)} = N(p)$, dann gilt

$$N(p) = \frac{K(p)}{l(p)} \leqq \frac{K(p) + Z}{l(p) + 1} = N(px) .$$

Es bleibt zu zeigen, daß $K(px) \geqq K(p) + N(p)$ erfüllt ist. Offenbar gilt $K_{\max}(p) \geqq$ $\geqq N(p)$. Aus Lemma 2 folgt aber, daß

$$K(px) \geqq K(p) + K_{\max}(p) .$$

b) Der Beweis erfolgt analog zu a).

Aus Satz 1 erhalten wir die Folgerungen:

a) $$N(\xi) = \lim_{n \to \infty} N(\xi[n]) ,$$

b) $$N(\eta) = \lim_{n \to \infty} N(\eta[-n, n]) .$$

Dabei sei ∞ als Grenzwert zugelassen.

Sei $\eta \in X^G$, dann soll $T\eta$ die um einen Takt nach links verschobene Folge η sein, d. h. $T\eta(i) =_{\mathrm{Df}} \eta(i + 1)$.

Lemma 3. *Sei $\eta \in X^G$, dann gilt $N(\eta) = N(T\eta)$.*

Beweis.

$$N(\eta[-n + 1, n - 1]) \leqq N(T\eta[-n, n]) \leqq N(\eta[-n - 1, n + 1]) .$$

Diese Ungleichungskette gilt nach der Folgerung 1 aus Satz 1 für $n \geqq 1$. Wenn wir in den einzelnen Komponenten zum Limes übergehen, erhalten wir

$$N(\eta) \leqq N(T\eta) \leqq N(\eta) ,$$

woraus die Behauptung folgt.

Die in dem letzten Lemma nachgewiesene Eigenschaft scheint uns generell wünschenswert zu sein für ein Maß, das die Komplexität der Struktur einer Folge aus X^G messen soll.

2. Charakterisierungssätze für Anfangsstück- bzw. Mittelstückmengen von Folgen

In diesem Abschnitt wollen wir einige Beziehungen zwischen der Komplexität von Folgen aus X^{N^+} und X^G und sprachtheoretischen Eigenschaften ihrer Anfangsstück- bzw. Mittelstückmengen aufzeigen.

Mit pq^{N^+} und q^G wollen wir die Folgen aus X^{N^+} bzw. X^G bezeichnen, die dadurch entstehen, daß wir an p „unendlich oft" q heranhängen bzw. vom 0-Punkt ausgehend q nach beiden Seiten „unendlich oft" hintereinanderschreiben.

Satz 2. *Sei $\xi \in X^{N^+}$ und $\eta \in X^G$; dann gilt:*

a) $N(\xi) < \infty \leftrightarrow \exists p \exists q (p, q \in X^* \wedge \xi = pq^{\mathbb{N}^+})$ [1]),

b) $N(\eta) < \infty \leftrightarrow \exists p (p \in X^* \wedge \eta = p^{G})$ [1]).

Beweis. a) ($\leftarrow$) Sei $\xi = pq^{\mathbb{N}^+}$. $N(\xi) = \lim_{i \to \infty} N(pq^i) < \infty$ ist zu zeigen. Für beliebige $j, i \in \mathbb{N}^+$ gilt aber:

$$K(pq^i, j) \leqq l(pq),$$

welches

$$N(\xi) \leqq l(pq)$$

und damit die Behauptung nach sich zieht.

($\rightarrow$) Sei $N(\xi) = C < \infty$. Dann muß

$$K(\xi[n], j) \leqq C$$

für alle $n, j \in \mathbb{N}^+$ erfüllt sein. Wenn dies nicht der Fall wäre, dann könnte $N(\xi) = C$ nicht gelten (Lemma 2b).

Hieraus kann man leicht ableiten, daß ξ die geforderte Gestalt haben muß.

b) Der Beweis erfolgt analog zu a).

Aus Satz 2 folgt, daß für eine Folge $\eta \in X^G$, deren Mittelstückmenge regulär ist,

$$\lim_{n \to \infty} \frac{N(\eta[-n, n])}{2n + 1} = \lim_{n \to \infty} \frac{K(\eta[-n, n])}{(2n + 1)^2} = 0$$

gilt.

Im nachfolgenden Satz 3 werden wir die Folgen charakterisieren, deren Mittelstückmengen kontextfrei sind, und eine notwendige Bedingung für das Wachstum der Funktion $K(\eta[-n, n])$ in Abhängigkeit von n für derartige Folgen angeben.

Sei $\xi \in X^{\mathbb{N}^+}$, dann verstehen wir unter $\overleftarrow{\xi}$ die folgende Funktion:

$$\overleftarrow{\xi}(-i) =_{\mathrm{Df}} \xi(i + 1) \quad \text{für alle } i \in \mathbb{N}.$$

Satz 3. *Sei $\eta \in X^G$; dann gilt:*

a) $\{\eta[-n, n] \mid n \in \mathbb{N}^+\}$ *ist kontextfrei*

$$\leftrightarrow \exists \xi_1 \exists \xi_2 (\xi_1, \xi_2 \in X^{\mathbb{N}^+} \wedge N(\xi_1) < \infty \wedge N(\xi_2) < \infty \wedge \eta = \overleftarrow{\xi}_1 \xi_2),$$

b) $\{\eta[-n, n] \mid n \in \mathbb{N}^+\}$ *sei kontextfrei; dann ist*

$$\lim_{n \to \infty} \frac{N(\eta[-n, n])}{2n + 1} \leqq \frac{1}{4}.$$

Beweis. a) ($\leftarrow$) Diese Implikation läßt sich mit der Konstruktion einer kontextfreien Grammatik G für die Menge $\{\eta[-n, n] \mid n \in \mathbb{N}^+\}$ nachweisen.

$G_1 = [X, H_1, \sigma_1, R_1]$ sei eine Grammatik für $\{\overleftarrow{\xi}_1[-n] \mid n \in \mathbb{N}^+\}$. Dabei sollen die Regeln r von R_1 nur die Formen

$$r = (h, xh') \quad \text{oder} \quad r = (h, x)$$

besitzen, wobei $h, h' \in H_1$ und $x \in X$ sind.

$G_2 = [X, H_2, \sigma_2, R_2]$ sei Grammatik für $\{\xi_2[n] \mid n \in \mathbb{N}^+\}$. Dabei sollen die Regeln r von R_2 nur die Formen

$$r = (h, h'x) \quad \text{oder} \quad r = (h, x)$$

besitzen, wobei $h, h' \in H_2$ und $x \in X$ sind.

---

[1]) Diese Darstellungen der Folgen sind jeweils genau dann möglich, wenn $\{\xi[n] \mid n \in \mathbb{N}^+\}$ bzw. $\{\eta[-n, n] \mid n \in \mathbb{N}^+\}$ regulär sind.

Diese Formen der Regeln lassen sich realisieren, da die betrachteten Mengen regulär sind.

Wir konstruier n nun die Grammatik $G = [X, H_1 \times H_2 \cup \{\sigma\}, \sigma, R]$, wobei für R gilt: $([h_1, h_2], p) \in R$ falls $h_1 \in H_1$, $h_2 \in H_2$ und p die Gestalt

$$p = x[h_1', h_2']\, y \quad \text{und} \quad (h_1, xh_1') \in R_1 \quad \text{und} \quad (h_2, h_2'y) \in R_2$$

oder

$$p = xy \quad \text{und} \quad (h_1, x) \in R_1 \quad \text{und} \quad (h_2, y) \in R_2$$

hat. Außerdem sind die Regeln $(\sigma, x[h_1, \sigma_2])$ in R, falls (σ_1, xh_1) in R_1 ist. Wie man leicht nachrechnet, leistet G das Verlangte.

($\rightarrow$) Diese Implikation folgt unter Anwendung des „*xuwvy*-Lemmas" (Lemma 3.1.1 in [7]) für kontextfreie Sprachen in wenigen Schritten und wird daher weggelassen.

b) Wenn $\{\eta[-n, n] \mid n \in \mathbb{N}^+\}$ kontextfrei ist, so wissen wir aus Satz 2 und Satz 3a, daß η die Form $\eta = p-\mathbb{N}^+qr\mathbb{N}^+$ haben muß.

Sei $C = K(pqr)$, dann läßt sich $K(p^vqr^v)$ nach oben abschätzen durch

$$K(p^vqr^v) \leqq C + (v-1)\,(i^2 + k^2 + ji + kj) + (v^2 + 2v)\, ki\,.$$

Dabei ist $i = l(p)$, $j = l(q)$ und $k = l(r)$. Wir dividieren diese Ungleichung durch $l(p^vqr^v)^2$ und gehen zum Limes für v gegen ∞ über:

$$\lim_{v\to\infty} \frac{K(p^vqr^v)}{l(p^vqr^v)^2} \leqq \lim_{v\to\infty} \frac{v^2ki}{v^2(i^2 + k^2 + 2ki)} \leqq \lim_{v\to\infty} \frac{v^2ki}{4v^2ki} = \frac{1}{4}\,.$$

Die letzte Ungleichung erhalten wir, wenn wir die Beziehung $i^2 + k^2 \geqq 2ki$ ausnutzen.

Beispiel. Sei $\eta = a-\mathbb{N}^+b\mathbb{N}$. Die Menge $\{a^{n+1}b^n \mid n \in \mathbb{N}^+\}$ ist kontextfrei. Für diese Folge gilt

$$\lim_{n\to\infty} \frac{N(a^{n+1}b^n)}{2n+1} = \frac{1}{4}\,,$$

woraus folgt, daß sich die Bedingung von Satz 3b nicht mehr verschärfen läßt.

Der Satz 3a wurde nur für Folgen aus X^G formuliert, weil für Folgen ξ aus $X^{\mathbb{N}^+}$ die Beziehung gilt:

Die Wortmenge $\{\xi[n] \mid n \in \mathbb{N}^+\}$ ist genau dann kontextfrei,
wenn sie regulär ist.

Die Aussage des Satzes 3b können wir in nachstehender Weise verallgemeinern. $L \subseteqq X^*$ sei eine kontextfreie Sprache mit $|L \cap X^n| \leqq 1$ für alle $n \in \mathbb{N}^+$. Dann erfüllt jede Folge $(p_i)_{i\in\mathbb{N}^+} \subseteqq L$ mit $l(p_i) < l(p_{i+1})$ für alle i die Ungleichung:

$$\lim_{i\to\infty} \frac{K(p_i)}{l(p_i)^2} \leqq \frac{1}{4}\,.$$

Diese Aussage hängt nicht von der Größe des Alphabetes X ab.

Wir wollen nun kontextsensitive Sprachen untersuchen, die pro Länge höchstens ein Wort enthalten.

Sei also $L \subseteqq X^*$ eine beliebige kontextsensitive Sprache mit $|L \cap X^n| \leqq 1$ für alle $n \in \mathbb{N}^+$. Dann ergibt sich die Frage, ob es ein c mit $1/4 \leqq c < 1/2$ gibt, so daß für jede Folge $(p_i)_{i\in\mathbb{N}^+} \subseteqq L$ mit $l(p_i) \leqq l(p_{i+1})$ für alle i gilt:

$$\limsup_{i\to\infty} \frac{K(p_i)}{l(p_i)^2} \leqq c\,.$$

Hierauf können wir folgende Teilantwort geben.

Satz 4. *Für jedes* $\varepsilon > 0$ *existiert ein Alphabet* X *und eine kontextsensitive Sprache* $L \subseteq X^*$ *mit* $|L \cap X^n| \leqq 1$ *für alle* $n \in \mathbb{N}^+$, *so daß für alle Folgen* $(p_i)_{i \in \mathbb{N}^+} \subseteq L$ *mit* $l(p_i) \leqq l(p_{i+1})$ *für alle* i *gilt:*

$$\lim_{i\to\infty} \frac{K(p_i)}{l(p_i)^2} > \frac{1}{2} - \varepsilon .$$

Beweis. Sei $X_i =_{\mathrm{Df}} \{1, 2, \ldots, i\}$, $M_i =_{\mathrm{Df}} \{1^j 2^j \cdots i^j / j \in \mathbb{N}^+\}$ und $p(i, j) =_{\mathrm{Df}} 1^j 2^j \cdots i^j$ dann gilt:

a) M_i ist kontextsensitiv für alle i (dies setzen wir als bekannt voraus),

b) $$\lim_{j\to\infty} \frac{K(p(i,j))}{l(p(i,j))^2} = \frac{1}{2} - \frac{1}{2i} .$$

Wir wollen dazu $K(p(i, j))$ berechnen. Es gibt offensichtlich i Teilwörter der Länge 1 in $p(i, j)$. Wenn die Länge k der Teilwörter im Intervall $2 \leqq k < j$ liegt, so ist die Anzahl der verschiedenen Teilwörter jeweils $ki - k + 1$. Wenn die Länge k der Teilwörter größer oder gleich j ist, so sind sämtliche Teilwörter verschieden. Daraus ergibt sich:

$$K(p(i, j)) = i + \sum_{k=2}^{j-1} (ki - k + 1) + \sum_{k=1}^{ji-j+1} k .$$

Dies kann vereinfacht werden, und man erhält

$$K(p(i, j)) = \frac{j^2 i^2 - ij^2 + 2ij}{2} .$$

Nun kann man durch $l(p(i, j))^2$ dividieren und zum Limes übergehen:

$$\lim_{j\to\infty} \frac{K(p(i,j))}{l(p(i,j))^2} = \lim_{l\to\infty} \frac{i^2 j^2 - ij^2 + 2ij}{2i^2 j^2} = \frac{1}{2} - \frac{1}{2i} .$$

Aus a) und b) folgt die Behauptung des Satzes. Man muß nur $i > 1/2\varepsilon$ wählen, dann erfüllen alle Folgen aus $M_i \subseteq X_i^*$ mit der im Satz formulierten Eigenschaft das von ihnen Verlangte.

Mit diesem Satz wurde nachgewiesen, daß es zumindest kein universelles c (unabhängig von der Größe des Alphabetes X) geben kann.

3. Die Regularität der Schnittpunktsprachen M_c

Es werden folgende Mengen betrachtet:

$$M_c =_{\mathrm{Df}} \{p / p \in X^* \wedge N(p) \leqq c\} .$$

Dabei ist c eine beliebige reelle Zahl. Diese Wortmengen wollen wir Schnittpunktsprachen nennen, denn sie sind in ähnlicher Weise definiert wie die Schnittpunktsprachen in der Theorie der stochastischen Automaten (siehe [9]). Unter einer solchen Sprache versteht man in dieser Theorie eine Menge von Eingabewörtern, deren Wahrscheinlichkeit, einen stochastischen Automaten ausgehend von einem ausgezeichneten Anfangszustand in eine vorgegebene Endmenge zu überführen, kleiner oder gleich einer fixierten reellen Zahl ist. Dabei wird also der zu einem Wort p gehörige Wert mit Hilfe des Verhaltens des Automaten errechnet, in unserem Fall geschieht dies mit der Funktion $N(.)$. Wir werden nachweisen, daß sämtliche Schnittpunktsprachen M_c regulär sind. Dazu benötigen wir die folgenden Definitionen und Hilfssätze.

Definition. Sei $p \in X^*$.

a) (p_1, p_2, p_3) heißt *Darstellung* von $p =_{\mathrm{Df}}$

1) Es gibt ein $i \in N^+$ und ein $q \in X^*$ mit q ist echtes Anfangsstück von p_2 ($q \sqsubset p_2$) und $p = p_1 p_2^i q p_3$,

2) $p_1, p_2, p_3 \in X^*$.

b) Die Darstellung (p_1, p_2, p_3) von p heißt *optimal* $=_{\mathrm{Df}}$

1) (p_1, p_2, p_3) ist Darstellung von p,

2) wenn (q_1, q_2, q_3) Darstellung von p ist, so gilt

$$l(p_1 p_2 p_3) \leqq l(q_1 q_2 q_3) .$$

c) $M_{p_1 p_2 p_3} =_{\mathrm{Df}} \{p/p \in X^* \wedge (p_1, p_2, p_3) \text{ ist Darstellung von } p\}$.

Lemma 4. *Für* $p \in M_{p_1 p_2 p_3}$ *gilt* $N(p) \leqq l(p_1 p_2 p_3)$.

Beweis. $p \in M_{p_1 p_2 p_3}$ bedeutet, es gibt i und q mit $p = p_1 p_2^i q p_3$. Wir werden zeigen, daß für ein solches p $K(p, j) \leqq l(p_1 p_2 p_3)$ für alle $j \in N^+$ gilt. Daraus folgt unmittelbar die Behauptung. Im Bereich p_1 können höchstens $l(p_1)$ verschiedene Teilwörter der Länge j beginnen. Im Bereich $p_2^i q$ können höchstens $l(p_2)$ verschiedene Teilwörter der Länge j beginnen,die nicht in p_3 enden und schließlich können im Bereich von p_3 höchstens $l(p_3)$ verschiedene Teilwörter der Länge j enden. Daraus folgt aber, daß es höchstens $l(p_1) + l(p_2) + l(p_3) = l(p_1 p_2 p_3)$ verschiedene Wörter der Länge j in $p_1 p_2^i q p_3$ geben kann.

Lemma 5. *Sei* (p_1, p_2, p_3) *optimale Darstellung von* $p = p_1 p_2^i q p_3$; *dann gilt:*

a) (p_1, p_2, p_3) *ist auch optimale Darstellung von* $p' = p_1 p_2^{i+1} q p_3$;

b) $N(p) \leqq N(p')$;

c) $\lim\limits_{j \to \infty} N(p_1 p_2^{i+j} q p_3) = l(p_1 p_2 p_3)$.

Beweis. a) Sei dem Leser überlassen.

b) Zu zeigen ist:

$$\frac{K(p)}{l(p)} \leqq \frac{K(p')}{l(p) + l(p_2)} .$$

Dies ist aber genau dann der Fall, wenn

$$Z =_{\mathrm{Df}} K(p') - K(p) \geqq l(p_2)\, N(p) .$$

Da (p_1, p_2, p_3) optimale Darstellung von p ist, läßt sich

$$Z = l(p_2)\, l(p_1 p_2 p_3)$$

nachweisen. Mit Hilfe von Lemma 4 folgt daraus $Z \geqq l(p_2)\, N(p)$ und damit die Behauptung von b).

c) Aus b) wissen wir, daß für den Zuwachs Z immer $Z = l(p_2)\, l(p_1 p_2 p_3)$ gilt. Sei $C = K(p_1 p_2^i q p_3)$, dann ist

$$N(p_1 p_2^{i+j} q p_3) = \frac{C + j l(p_2)\, l(p_1 p_2 p_3)}{l(p_1 p_2^i q p_3) + j l(p_2)} .$$

Beim Übergang zum Limes ergibt sich dann:

$$\lim_{j \to \infty} N(p_1 p_2^{i+j} q p_3) = \lim_{j \to \infty} \frac{C + j l(p_2)\, l(p_1 p_2 p_3)}{l(p_1 p_2 p_3) + j l(p_2)} = \lim_{j \to \infty} \frac{j l(p_2)\, l(p_1 p_2 p_3)}{j l(p_2)} = l(p_1 p_2 p_3) .$$

Sei c eine natürliche Zahl und $K(p, c+1) \geqq c+1$. Wenn außerdem $n = l(p) > 2c^2 + 3c + 1$ gilt, so muß auch $N(p) > c$ sein. Man erhält dies durch folgende Überlegungen:

$$N(p) = \frac{K(p)}{l(p)} \geqq \frac{(n-2c-1)(c+1)}{n} = f(n)\,.$$

f ist streng monoton wachsend. Für $n = 2c^2 + 3c + 1$ ergibt sich

$$f(2c^2 + 3c + 1) = \frac{(2c^2 + c)(c+1)}{2c^2 + 3c + 1} = c\,.$$

Aus diesen Bemerkungen folgt, daß für ein Wort p mit $l(p) > 2c^2 + 3c + 1$ und $p \in M_c$ gelten muß: $K(p, c+1) \leqq c$.

Lemma 6. *Sei* $p \in X^*$ *mit* $l(p) > 2c^2 + 3c + 1$ *und* $K(p, c+1) \leqq c$; *dann gibt es eine Darstellung* (p_1, p_2, p_3) *von* p *mit* $l(p_1 p_2 p_3) < c$.

Beweis. Aus $K(p, c+1) \leqq c$ folgt mit Hilfe von Lemma 2, daß $K(p, j) \leqq c$ für alle $j \in \mathbb{N}^+$ gilt.

Angenommen (p_1, p_2, p_3) wäre eine optimale Darstellung von p mit $p = p_1 p_2^i q p_3$ und $l(p_1 p_2 p_3) > c$. Wie man leicht nachprüft, gibt es dann mindestens $c + 1$ verschiedene Teilwörter der Länge c in p, d. h. $K(p, c) \geqq c + 1$. Dies ist aber ein erwünschter Widerspruch zu unserer Annahme.

Satz 5. *c sei eine natürliche Zahl. Dann gilt:*

a) $M_c = \bigcup\limits_{\substack{p_1, p_2, p_3 \in X^* \\ l(p_1 p_2 p_3) \leqq c}} M_{p_1 p_2 p_3} \cup M$, *wobei* $M = M_c \cap X^{[2c^2+3c+1]}$ ist.

b) M_c *ist regulär.*

Beweis. a) folgt unmittelbar aus den Lemmata 4, 5 und 6.

b) $M_{p_1 p_2 p_3}$ sind reguläre Mengen und M ist sogar endlich. Somit ist M_c eine endliche Vereinigung von regulären Mengen und damit selbst regulär.

Satz 6. *c sei eine beliebige reelle Zahl. Dann gilt:*

a) $M_c = \bigcup\limits_{\substack{p_1, p_2, p_3 \in X^* \\ l(p_1 p_2 p_3) \leqq c}} M_{p_1 p_2 p_3} \cup M$, *wobei* $M \subseteq X^*$ *endlich ist.*

b) M_c *ist regulär.*

Beweis. a) Es gilt nachzuweisen, daß M endlich gewählt werden kann. Sei

$$L_c =_{\mathrm{Df}} \bigcup_{\substack{p_1, p_2, p_3 \in X^* \\ l(p_1 p_2 p_3) \leqq c}} M_{p_1 p_2 p_3}\,.$$

Wenn $p \in L_{c+1} - L_c$ und (p_1, p_2, p_3) optimale Darstellung von $p = p_1 p_2^i q p_3$ ist, dann gilt nach Lemma 5:

1) für alle $j \in \mathbb{N}^+$ ist $p_1 p_2^{i+j} q p_3 \in L_{c+1} - L_c$,

2) $\lim\limits_{j \to \infty} N(p_1 p_2^{i+j} q p_3) = l(p_1 p_2 p_3) = [c] + 1$.

Danach existiert ein n_0, so daß $N(p_1 p_2^k q p_3) > c$ für alle $k > n_0$. Für die $p \in L_{c+1} - L_c$ kann es nur endlich viele verschiedene optimale Darstellungen geben. Sei $n_{\max}$ das Maximum der zu diesen optimalen Darstellungen jeweils gehörenden n_0. Hiernach gilt:

$$(L_{c+1} - L_c) \cap M_c \subseteq X^{[(n_{\max}+3)([c]+1)]}\,.$$

Sei nun $m = \max\{(n_{\max} + 3)([c] + 1), 2([c] + 1)^2 + 3([c] + 1) + 1\}$. Dann kann M in der folgenden Weise gewählt werden:

$$M = X^{[m]} \cap M_c \,.$$

Der Beweis von b) erfolgt wie bei Satz 5b.

Die Umkehrung von Satz 6 gilt nicht, d. h., nicht jede reguläre Wortmenge läßt sich als Schnittpunktsprache M_c darstellen. Das folgende Beispiel dokumentiert dies: Sei $L = \{(ab)^n / n \in \mathbb{N}^+\}$ und sei c so gewählt, daß $L \subseteqq M_c$ gilt; dann ist aber auch $L' = \{a^n / n \in \mathbb{N}^+\}$ ganz in M_c enthalten.

4. Zum Verhalten der Funktionen *K* und *N* bzgl. Homomorphismen

In [1] gelang es, mit Hilfe der Teilwortkomplexität gewisse Typen von determinierten Lindenmayer-Sprachen zu charakterisieren. Daß sich diese enge Beziehung zwischen diesem Komplexitätsmaß und jenen Typen von Lindenmayer-Sprachen ergibt, liegt nach unserer Meinung vor allem an der simultanen Ersetzung beim Ableiten, die für die Lindenmayer-Systeme charakteristisch ist. Die simultane Ersetzung entspricht der Ausführung eines Homomorphismus, und daher wollen wir zum Abschluß dieser Arbeit einige Eigenschaften angeben, die die Funktionen K und N bzgl. Homomorphismen besitzen. Wir fassen diese in folgendem Satz zusammen.

Satz 7. *Seien X, Y endliche Alphabete. Durch $H'\colon X \to Y^*$ legen wir einen Homomorphismus H von X^* nach Y^* fest, wobei $H(e) =_{\mathrm{Df}} e$ und $H(px) =_{\mathrm{Df}} H(p)\,H'(x)$ für $p \in X^*$ und $x \in X$ sein soll.*

Dann gelten folgende Aussagen:

a) *Wenn $H'\colon X \to Y$, so gilt für alle $p \in X^*$:*

(1) $K(p) \geqq K(H(p))$,

(2) $N(p) \geqq N(H(p))$.

Die Gleichheit gilt für alle $p \in X^*$ genau dann, wenn H' zusätzlich noch eineindeutig ist.

b) *$H\colon X^* \to Y^*$ ist eineindeutig genau dann, wenn für alle $p \in X^*$ gilt:*

$$K(p) \leqq K(H(p)) \,.$$

c) *Wenn $N(p) \leqq N(H(p))$ für alle $p \in X^*$ gilt, so ist H eineindeutig.*

d) *Wenn $H'\colon X \to Y^l$ eineindeutig ist, so gilt für alle $p \in X^*$: $N(p) \leqq N(H(p))$. Dies ist ein Spezialfall für die Umkehrung von* c).

Beweis: a) Ist unmittelbar einzusehen.

b) ($\leftarrow$) Wir zeigen die Kontraposition. $\{H(x)/x \in X\}$ ist nicht frei; dann gibt es ein $p \in X^*$ mit $K(p) > K(H(p))$.

Wenn $\{H(x)/x \in X\}$ nicht frei ist, dann existiert ein $p_0 \in Y^*$ mit zwei Darstellungen aus $\{H(x)/x \in X\}$. Wir bilden zwei Folgen $(p_i)_{i \in \mathbb{N}^+}$ und $(q_i)_{i \in \mathbb{N}^+}$, wobei

$$p_i = \underbrace{p_0 p_0 \cdots p_0}_{i\text{-mal}}\,, \qquad q_i = \underbrace{q_0^1 q_0^2 q_0^1 q_0^1 q_0^2 q_0^1 q_0^1 q_0^1 q_0^2 \cdots}_{i\text{-mal}}\,,$$

$H(q_0^1) = H(q_0^2) = p_0$ und $q_0^1 \neq q_0^2$ gelten soll. Der Zusammenhang zwischen p_i und q_i ergibt sich dann durch die Gleichung $H(p_i) = H(q_i)$ für alle $i \in \mathbb{N}^+$.

Nach Lemma 4 gilt $N(p_i) \leqq l(p_0)$ für alle $i \in \mathbb{N}^+$. Sei nun $\xi = q_0^1 q_0^2 q_0^1 q_0^1 q_0^2 q_0^1 q_0^1 q_0^1 q_0^2 q_0^1 \ldots$ die durch dieses Bildungsgesetz entstehende Folge aus $X^{\mathbb{N}^+}$. Es läßt sich leicht nachweisen, daß die Anfangsstückmenge von ξ nicht regulär sein kann. Daraus und aus Satz 2 folgt damit, daß es für alle $M \in N^+$ ein n_0 geben muß, so daß $N(q_{n_0}) > M$ gilt. Somit wächst $K(p_i)$ linear mit i, hingegen wächst $K(q_i)$ aber stärker als linear mit i. Damit ist die Existenz eines p mit $K(p) > K(H(p))$, wie oben behauptet, nachgewiesen.

($\rightarrow$) $\{H(x)/x \in X\}$ sei frei. p_1, p_2 seien Teilwörter von p und p_1 sei verschieden von p_2. Dann sind aber auch $H(p_1)$ und $H(p_2)$ verschieden und Teilwörter von $H(p)$. Somit ist $K(p) \leqq K(H(p))$.

c) Wir weisen wieder die Kontraposition nach. $\{H(x)/x \in X\}$ sei nicht frei; dann existiert ein $p \in X^*$ mit $N(p) > N(H(p))$. b) gewährleistet, daß es unter diesen Voraussetzungen zumindest ein p' geben muß mit $K(p') > K(H(p'))$. Da $l(p') \leqq l(H(p'))$ erfüllt ist, gilt für p' auch $N(p') > N(H(p'))$, womit die Behauptung bewiesen wäre.

d) Sei $p \in X^n$. Aus den an H gestellten Bedingungen folgt dann:

$$K(p, i) \leqq K(H(p), li) \quad \text{für alle } i \in \{1, \dots, n\} .$$

Weiter gilt:

$$K(H(p), li + j) \geqq K(p, i)$$

für alle $i \in \{1, \dots, n-1\}$ und alle $j \in \{0, \dots, n-1\}$. Daraus erhält man:

$$l \cdot \sum_{i=1}^{n-1} K(p, i) \leqq \sum_{i=l}^{ln-1} K(H(p), i) .$$

Zusammen mit der Ungleichung

$$l \leqq \sum_{i=1}^{l-1} K(H(p), i) + 1$$

ergibt das:

$$l \cdot \sum_{i=1}^{n} K(p, i) \leqq \sum_{i=1}^{nl} K(H(p), i) .$$

Diese Ungleichung dividieren wir durch nl, und wir erhalten die Behauptung

$$N(p) = \frac{\sum_{i=1}^{n} K(p, i)}{n} \leqq \frac{\sum_{i=1}^{nl} K(H(p), i)}{nl} = N(H(p)) .$$

Wir vermuten, daß die Umkehrung von c) allgemein gilt.

Literatur

[1] *Ehrenfeucht, A., K. P. Lee, G. Rozenberg*, Subword complexity of various classes of deterministic developmental languages without interactions. Theoret. Comp. Sci. **1** (1975), 59–75.

[2] *Колмогоров А. Н.*, Три подхода к определению понятия ,,количество информации''. Пробл. передачи информации **1** (1965) 1, 3—7.

[3] *Колмогоров А. Н.*, Несколько теорем об алгоритмическом количестве информации. УМН **23** (1968) 2, стр. 210.

[4] *Thiele, H.*, Einige Bemerkungen zur Weiterentwicklung der Informationstheorie. Nova acta Leopold. NF **27/1** (1972) Nr. 206.

[5] *Thiele, H.*, Zur Definition von Kompliziertheitsmaßen für endliche Objekte. In: Organismische Informationsverarbeitung (Herausgeber: F. Klix); Berlin 1974.

[6] *Böhm, H.*, Das Gittermodell. Diplomarbeit, Sektion Psychologie der Humboldt-Universität zu Berlin, 1970.

[7] *Ginsburg, S.*, The mathematical theory of context-free languages. New York 1966.
[8] *Lempel, A., J. Zif*, On the complexity of finite sequences. IEEE Trans. Inform. Theory **IT 22** (1976) 1.
[9] *Turakainen, P.*, On nonregular events representable on probabilistic automata with one input letter. Ann. Univ. Turku, Ser. A. I, **90** (1966), 3—14.

Kurzfassung

In der vorliegenden Arbeit wird die Teilwortkomplexität von Wörtern und Folgen über einem endlichen Alphabet X untersucht, und es werden einige Eigenschaften dieser Komplexität in bezug auf die Theorie formaler Sprachen nachgewiesen (z.B.: Die Eigenschaft eines Wortes p, eine Teilwortkomplexität kleiner oder gleich $c \cdot l(p)$ zu besitzen, wobei c eine fixierte reelle Zahl und $l(p)$ die Länge von p ist, ist regulär).

Abstract

In this paper the subword complexity of strings and sequences over a finite alphabet X is considered and some properties of this complexity in relation to the theory of formal languages are shown (e.g.: The regularity of the property of a string p to have a subword complexity less or equal to $c \cdot l(p)$, where c is a fixed real number and $l(p)$ is the length of p).

Резюме

В данной работе исследуется мера сложности слов и последовательностей над конечным алфавитом X. Как мера сложности используется число подслов данного слова. Доказываются некоторые свойства этой сложности относительно теории формальных языков (например: множество всех слов p, сложность которых меньше или равно $c \cdot l(p)$, причём c — фиксированное действительное число и $l(p)$ — длина слова p, регулярно).

(Eingang: erste Fassung am 5. 11. 1975,
vorliegende Fassung am 11. 6. 1976)

Anschrift des Verfassers:
Dipl.-Math. M. Heinz
102 Berlin
Weinmeisterstr. 12
DDR

Elektronische Informationsverarbeitung und Kybernetik
EIK 13 (1977) 1/2, 39–53

Über lineare Simulierbarkeit endlicher Automaten

Von *Hermann K.-G. Walter*

0. Einleitung

Bekanntermaßen kann über den Simulationsbegriff eine natürliche Verbindung zwischen Halbgruppentheorie und Theorie endlicher Automaten hergestellt werden. Dieser Zusammenhang ist zuerst entdeckt worden von K. Krohn und J. Rhodes [11], [1]. Verwendet man diesen Begriff in der von M. Arbib angegebenen Version [1], so kann wie im folgenden dargelegt wird, eine Äquivalenztheorie entwickelt werden, die weitgehend der klassischen Äquivalenztheorie entspricht. Hierdurch ergibt sich die Möglichkeit, Probleme der linearen Realisierung von Automaten mit halbgruppentheoretischen Methoden zu untersuchen.

Wir leiten eine Charakterisierung her, die in Verallgemeinerung von Ergebnissen aus [2], [3], [4], [8], [14], [15] eine vollständig halbgruppentheoretische Charakterisierung der linearen Realisierbarkeit von Automaten ergibt. Dies bedeutet, daß für Realisierungsfragen wesentliche Eigenschaften eines Automaten ausgedrückt werden können durch Eigenschaften der Transitionshalbgruppe.

Zur Frage der verschiedenen Realisierungsdefinitionen vergleiche man [9].

1. Bezeichnungen und Definitionen (vgl. [10])

Ein *endlicher Automat* a ist ein 5-tupel $a = (I(a), O(a), S(a), \delta_a, \lambda_a)$, wobei die $j(a)$ endliche Mengen für $j = I, O, S$ und

$$\delta_a\colon I(a) \times S(a) \to S(a) \;\&\; \lambda_a\colon I(a) \times S(a) \to O(a)$$

Abbildungen sind.

Wir wollen im folgenden stets annehmen, daß λ_a surjektiv ist, daß also keine überflüssigen Ausgabesymbole existieren.

Wir setzen δ_a und λ_a auf $I(a)^* \times S(a)$ fort gemäß den Formeln

i) $\lambda_a(wv, s) = \lambda_a(w, s), \lambda_a(v, \delta_a(w, s))$

ii) $\delta_a(wv, s) = \delta_a(v, \delta_a(w, s))$

für $w, v \in I(a)^*$ und $s \in S(a)$. Dabei ist $I(a)^*$ das freie Monoid über $I(a)$.

Seien a und a' zwei Automaten. Ein Tripel $\varphi = (\varphi_I, \varphi_O, \varphi_S)$ von Abbildungen

$$\varphi_j\colon j(a) \to j(a') \quad (j = I, O, S)$$

heißt *Homomorphismus*, (in Zeichen $\varphi\colon a \to a'$), falls die Diagramme

$$\text{(H1)}\quad \begin{array}{ccccc} I(a) & \times & S(a) & \xrightarrow{\delta_a} & S(a) \\ \varphi_I\downarrow & & \downarrow\varphi_S & & \downarrow\varphi_S \\ I(a') & \times & S(a') & \xrightarrow[\delta_{a'}]{} & S(a') \end{array}$$

und

$$\text{(H2)}\quad \begin{array}{ccccc} I(a) & \times & S(a) & \overset{\lambda_a}{\rightarrow} & O(a) \\ \varphi_I\downarrow & & \downarrow\varphi_S & & \downarrow\varphi_O \\ I(a') & \times & S(a') & \underset{\lambda_{a'}}{\rightarrow} & O(a') \end{array}$$

kommutativ sind.

Falls nur (H1) kommutativ ist, sprechen wir von *δ-Homomorphie* (in Zeichen: $\varphi: a \overset{\delta}{\rightarrow} a'$). *Monomorphismen* ($a \underset{\varphi}{\leqq} a'$ oder $a \leqq a'$), *Epimorphismen* und *Isomorphismen* ($a \underset{\varphi}{\triangleq} a'$ oder $a \triangleq a'$) werden wie üblich definiert. Sind φ_I und φ_O Identitäten, so sprechen wir von *S-Homomorphismen*, entsprechend von *(I, S)-Homomorphismen*, etc. S-Epimorphismen heißen *Reduktionen*.

Die Konzepte übertragen sich soweit sinnvoll auf δ-Homomorphismen. Insbesondere schreiben wir

$$\text{„}a \overset{(\delta)}{\triangleq} a'\text{“} \quad \text{und} \quad \text{„}a \overset{(\delta)}{\leqq} a'\text{“}.$$

Unterautomaten und *δ-Unterautomaten* sind wie üblich definiert. Wir schreiben „$a \subseteq a'$“ und „$a \overset{(\delta)}{\subseteq} a'$“.

Seien a und a' Automaten mit $I(a) = I(a')$, $O(a) = O(a')$, seien ferner $s \in S(a)$ & & $s' \in S(a')$, s und s' sind *äquivalent* (in Zeichen: $s \sim s'$), falls

$$\lambda_a(w, s) = \lambda_{a'}(w, s')$$

für alle $w \in I(a)^*$ gilt.

Die Menge

$$F(a) = \{\lambda_a^s \mid \lambda_a^s(w) = \lambda_a(w, s),\ s \in S(a),\ w \in I(a)^*\}$$

ist die zu a gehörige *Abbildungsfamilie*.

Zwei Automaten a und a' heißen *funktional äquivalent* (in Zeichen: $a \sim a'$), falls $F(a) = F(a')$ gilt.

Ein Automat a heißt *reduziert*, falls je zwei Zustände inäquivalent sind.

Bekanntermaßen definiert die Äquivalenz von Zuständen eine „S-Kongruenz“ und damit einen Automaten $\mathrm{Red}_S(a)$. Für diesen gilt:

1) Es existiert eine Reduktion $\varrho: a \rightarrow \mathrm{Red}_S(a)$
2) $a \sim \mathrm{Red}_S(a)$
3) $a \sim a' \Leftrightarrow \mathrm{Red}_S(a) = \mathrm{Red}_S(a')$
4) $\mathrm{Red}_S(a)$ ist reduziert.

Definition 1.1 ([1]). Seien a und a' Automaten. a *simuliert* a' (in Zeichen: a'/a), falls es Abbildungen

$$h_I: I(a') \rightarrow I(a)\,, \quad h_O: O(a) \rightarrow O(a')\,, \quad h_S: S(a') \rightarrow S(a)$$

gibt, so daß für alle $w \in I(a')^*$ und $s \in S(a')$

$$\lambda_{a'}(w, s) = h_O\big(\lambda_a(h_I(w), h_S(s))\big)$$

gilt. Dabei sind die homomorphen Fortsetzungen von h_I bzw. h_O wieder mit h_I bzw. h_O bezeichnet.

Eine Simulation erfolgt so, daß alle Rechenarbeit von a' durch a erfolgt und h_I und h_O nur Kodierer und Dekodierer sind.

Abb. 1 veranschaulicht den Begriff der Simulation.

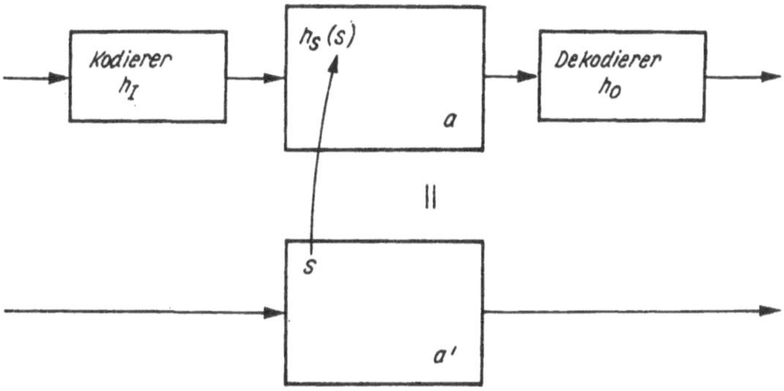

Abb. 1

Ein Tripel h, wie es in der Definition verlangt wird, heißt *Simulation* (in Zeichen: $h: a' \Rrightarrow a$).

Folgende Beobachtung, deren einfachen Beweis wir übergehen, zeigt, daß Simulationen Verallgemeinerungen der δ-Homomorphie sind.

Beobachtung. *Sei $h: a \Rrightarrow a'$ eine Simulation; dann gilt für alle $u, w \in I(a)^*$ und $s \in S(a)$*

$$\lambda_a(u, \delta a(w, s)) = h_0(\lambda_{a'}(h_I(u), \delta a'(h_I(w), h_S(s)))) .$$

Bemerkung. Es gilt stets

1) a/a

2) a/a' & $a'/a'' \Rightarrow a/a''$.

Zwei Automaten a und a' heißen *schwach äquivalent* (in Zeichen: $a \approx a'$), falls a/a' und a'/a gilt.

Bemerkung. Es gilt stets

1) $a \approx a$,

2) $a \approx a'$ & $a' \approx a'' \Rightarrow a \approx a''$,

3) $a \approx a' \Rightarrow a' \approx a$.

Wir beschließen diesen Abschnitt mit der Angabe zweier Operationen zwischen Automaten und zwei Klassen von Automaten, die wir im folgenden benötigen.

Seien a_1 und a_2 zwei Automaten mit $I(a_1) = I(a_2)$ und $S(a_1) \cap S(a_2) = \varnothing$. Dann definiere $a_1 \oplus a_2$ durch

1) $I(a_1 \oplus a_2) = I(a_1)$,

$O(a_1 \oplus a_2) = O(a_1) \cup O(a_2)$,

$S(a_1 \oplus a_2) = S(a_1) \cup S(a_2)$.

2) $$\delta_{a_1 \oplus a_2}(x, s) = \begin{cases} \delta_{a_1}(x, s) & \text{für } s \in S(a_1)\,, \\ \delta_{a_2}(x, s) & \text{für } s \in S(a_2)\,; \end{cases}$$

$$\lambda_{a_1 \oplus a_2}(x, s) = \begin{cases} \lambda_{a_1}(x, s) & \text{für } s \in S(a_1)\,, \\ \lambda_{a_2}(s, s) & \text{für } s \in S(a_2) \end{cases}$$

$(x \in I(a))$.

Für die zweite Operation setzen wir nun voraus $I(a_1) = I(a_2)$. Dann definiere $a_1 \otimes a_2$ durch

1) $I(a_1 \otimes a_2) = I(a_1)$, $S(a_1 \otimes a_2) = S(a_1) \times S(a_2)$,
$O(a_1 \otimes a_2) = O(a_1) \times O(a_2)$

2) $\delta_{a_1 \otimes a_2}(x, s_1, s_2) = (\delta_{a_1}(x, s_1), \delta_{a_2}(x, s_2))$,
$\lambda_{a_1 \otimes a_2}(x, s_1, s_2) = (\lambda_{a_1}(x, s_1), \lambda_{a_2}(x, s_2))$
$(x \in I(a_1), s_1 \in S(a_1) \;\&\; s_2 \in S(a_2))$.

Sei X eine Menge. Betrachte ein Symbol \$. Definiere den *Punktautomaten* 1_X durch

1) $I(1_X) = X$, $S(1_X) = \$$, $O(1_X) = \$$

2) $\delta_{1_X}(x, \$) = \lambda_{1_X}(x, \$) = \$ \; (x \in X)$.

Wir nennen einen Automaten a *zyklisch*, falls gilt: Es existiert ein $s \in S(a)$ mit $\delta_a(I(a)^*, s) = S(a)$.

2. Simulationen und schwache Äquivalenz von Automaten

Wir wollen Simulationen und schwache Äquivalenz von Automaten charakterisieren. Insbesondere wollen wir einen Zerlegungssatz herleiten. Hierzu benötigen wir das Konzept des *vollreduzierten Automaten* ([12], [13]).

Ein reduzierter Automat heißt *vollreduziert*, falls aus $\delta_a(x, s) = \delta_a(x', s) \;\&\; \lambda_a(x, s) = \lambda_a(x', s)$ für alle $s \in S(a)$ stets folgt: $x = x'$.

Zu jedem Automaten a können wir einen vollreduzierten Automaten $\mathrm{Red}(a)$ in folgender Weise konstruieren. Betrachte die Äquivalenzrelation

$$x \sim x' \Leftrightarrow \forall s \in S(a)\colon \delta_a(x, s) \sim \delta_a(x', s) \;\&\; \lambda_a(x, s) = \lambda_a(x', s) .$$

Dann wird durch

1) $I(\mathrm{Red}(a)) = I(a)/\sim$, $O(\mathrm{Red}(a)) = O(a)$, $S(\mathrm{Red}(a)) = S(\mathrm{Red}_S(a))$

2) $\delta_{\mathrm{Red}(a)}([x]_\sim, [s]_\sim) = [\delta_a(x, s)]_\sim \;\&\; \lambda_{\mathrm{Red}(a)}([x]_\sim, [s]_\sim) = \lambda_a(x, s)$

ein vollreduzierter Automat $\mathrm{Red}(a)$ definiert.

Lemma 2.1. *Seien a und a' Automaten; dann gilt:*

(1) $a/a' \Leftrightarrow \mathrm{Red}_S(a)/\mathrm{Red}_S(a') \Leftrightarrow \mathrm{Red}(a)/\mathrm{Red}(a')$;

(2) $a \approx \mathrm{Red}_S(a) \approx \mathrm{Red}(a)$.

Beweis. (1) folgt aus (2). Aus der Existenz der Reduktion $\varrho\colon a \to \mathrm{Red}_S(a)$ und der I-Epimorphismen $\tau\colon \mathrm{Red}_S(a) \to \mathrm{Red}(a)$ folgt $a/\mathrm{Red}_S(a)$ und $\mathrm{Red}_S(a)/\mathrm{Red}(a)$.

Wir haben zu zeigen:

$\mathrm{Red}(a)/a$. Sei $\varrho_v\colon a \to \mathrm{Red}(a)$ der kanonische (I, S)-Epimorphismus. Wähle h_I und h_S rechtsinvers zu $(\varrho_v)_I$ bzw. $(\varrho_v)_S$ und $h_O = 1_{O(a)}$. Dann gilt stets für $x \in I(\mathrm{Red}(a))$ und $s \in S(\mathrm{Red}(a))$:

$$\delta_a(h_I(x) , \quad h_S(s)) \sim h_S(\delta_{\mathrm{Red}(a)}(x, s)) .$$

Hieraus folgt aber sofort, daß $h = (h_I, h_S, h_O)$ eine Simulation ist.

Wir können nun die schwache Äquivalenz charakterisieren. Hierzu benötigen wir zwei einfache Lemmata.

Lemma 2.2. *Sei $h\colon a \Rightarrow a'$ eine Simulation und a vollreduziert, dann sind h_I und h_S injektiv.*

Beweis. 1. „h_S ist injektiv": Sei $h_S(s_1) = h_S(s_2)$ $(s_1, s_2 \in S(a))$, dann gilt für alle $w \in I(a)^*$

$$\lambda_a(w, s_1) = h_O(\lambda a'(h_I(w), h_S(s_1))) = h_O(\lambda a'(h_I(w), h_S(s_2))) = \lambda_a(w, s_2) .$$

Also gilt: $s_1 \sim s_2$. Da a reduziert, folgt $s_1 = s_2$.

2. „h_I ist injektiv“: Sei $h_I(x_1) = h_I(x_2)$ $(x_1, x_2 \in I(a))$. Dann gilt für alle $s \in S(a)$

$$\lambda_a(x_1, s) = h_O(\lambda a'(h_I(x_1), h_S(s))) = h_O(\lambda a'(h_I(x_2), h_S(s))) = \lambda_a(x_2 s) .$$

Ferner gilt für alle $s \in S(a)$ und $w \in I(a)^*$

$$\begin{aligned}\lambda_a(w, \delta_a(x_1, s)) &= h_O(\lambda a'(h_I(w), \delta_{a'}(h_I(x_1), h_S(s))))\\ &= h_O(\lambda a'(h_I(w), \delta_{a'}(h_I(x_2), h_S(s))))\\ &= \lambda_a(w, \delta_a(x_2, s)) ,\end{aligned}$$

also $\delta_a(x_1, s) \sim \delta_a(x_2, s)$. Hieraus folgt $x_1 = x_2$.

Lemma 2.3. *Sei $h: a \Rrightarrow a'$ eine Simulation, sei a ferner vollreduziert. Dann ist (h_I, h_S) ein δ-Homomorphismus, falls gilt:*

$$\delta a(h_I(I(a)^*) , \quad h_S(S(a))) \subseteqq h_S(S(a)) .$$

Beweis. Seien $x \in I(a)$ und $s \in S(a)$. Wähle s' so, daß gilt: $h_S(s') = \delta_{a'}(h_I(x), h_S(s))$. Dann gilt für alle $w \in I(a)^*$

$$\begin{aligned}&\lambda_a(w, s') = h_O(\lambda a(h_I(w), h_S(s'))) =\\ &= h_O(\lambda a(h_I(w), \delta_{a'}(h_I(x), h_S(s)))) =\\ &= \lambda_a(w, \delta_a(x, s)) .\end{aligned}$$

Also folgt: $s' \sim \delta_a(x, s)$. Hieraus folgt aber die Behauptung.

Theorem 2.1. *Seien a und a' Automaten; dann sind folgende Aussagen äquivalent:*-

1) $a \approx a'$

2) $\mathrm{Red}(a) \triangleq \mathrm{Red}(a')$.

Beweis. Wir haben zu zeigen: 1) $\Rightarrow$ 2). Ohne Beschränkung der Allgemeinheit können wir annehmen, daß a und a' vollreduziert sind. Sind dann $h: a \Rrightarrow a'$ und $h': a' \Rrightarrow a$ die zugehörigen Simulationen, so sind nach Lemma 2.2 h_I, h_S, h'_I, h'_S injektiv, also bijektiv. Dies bedeutet aber nach Lemma 2.3: (h_I, h_S) ist ein δ-Isomorphismus. Es verbleibt zu zeigen, daß h_O und h'_O surjektiv sind. Hieraus folgt h_O ist bijektiv und (h_I, h_S, h_O^{-1}) ist ein Isomorphismus. Sei $y = O(a)$, dann existiert $x \in I(a)$ und $s \in S(a)$ mit

$$y = \lambda_a(x, s) = h_O(\lambda a'(h_I(x), h_S(s))) .$$

Dies beweist aber die Surjektivität von h_O. Entsprechend folgt die Surjektivität von h'_O. Damit ist alles bewiesen.

Korollar. *Es gilt*

$$a \sim a' \Rightarrow a \approx a' .$$

Wir geben eine weitere Charakterisierung von Simulationen. Wir nennen einen Automaten a *ausgabefrei*, falls $\lambda_a(x, s) = (x, s)$ gilt $(x \in I(a), s \in S(a))$.

Lemma 2.4. *Sei a ein vollreduzierter und a' ein ausgabefreier Automat, dann simuliert a' genau dann a, wenn es ein Paar (h_I, h_S) von injektiven Abbildungen h_j: $j(a) \to j(a')$ $(j = I, S)$ gibt, so daß für alle $w, w' \in I(a)^*$ und $s, s' \in S(a)$ stets gilt:*

$$\delta_{a'}(h_I(w), h_S(s)) = \delta_{a'}(h_I(w'), h_S(s')) \Rightarrow \delta_a(w, s) = \delta_a(w', s') .$$

Beweis. ($\Rightarrow$). Sei $h: a \Rrightarrow a'$ eine zugehörige Simulation. Dann gilt für alle $u \in I(a)^*$

$$\begin{aligned}\lambda_a(u, \delta_a(w, s)) &= h_O(\lambda_{a'}(h_I(u), \delta a'(h_I(w), h_S(s))))\\ &= h_O(\lambda_{a'}(h_I(u), \delta_{a'}(h_I(w), h_S(s))))\\ &= \lambda_a(u, \delta_a(w's')) .\end{aligned}$$

Also folgt mit der Vollreduziertheit von a

$$\delta_a(w, s) = \delta_a(w', s') .$$

($\Leftarrow$). Definiere $h_O\colon O(a') \to O(a)$ durch

$$h_O(x, s) = \begin{cases} \lambda_a(x', \delta_a(w, s')) , & \text{falls } x = h_I(x') \text{ und} \\ & s = \delta_{a'}(h_I(w), h_S(s')) , \\ \text{beliebig} & \text{sonst} \end{cases}$$

$(x \in I(a'), s \in S(a'))$. Nach Voraussetzung ist h_O wohldefiniert. Ein einfacher Induktionsbeweis zeigt nun, daß (h_I, h_O, h_S) eine Simulation ist.

Wir ziehen einige Folgerungen.

Korollar 1. $a \overset{(\delta)}{\leqq} a' \Rightarrow a/a'$.

Bezeichnung. $\delta_a^w(s) = \delta_a(w, s)$ $(w \in I(a)^*, s \in S(a))$.

Korollar 2. *Ist* $h\colon a \Rrightarrow a'$ *eine Simulation und* a *vollreduziert, so gilt:*

$$\delta_a^{h_I(w)} = \delta_a^{h_I(w')} \Rightarrow \delta_a^w = \delta_a^{w'} .$$

Durch dieses Korollar gewinnen wir Anschluß an die Teilbarkeitstheorie von Halbgruppen [1]. Wir verallgemeinern geringfügig.

Ein Paar $[M, I]$ ist eine *Halbgruppe mit Generatornamen*, falls M eine Halbgruppe und I eine Menge ist, wobei es eine Namensfunktion $\beta\colon I \to M$ gibt, so daß $\beta(I)$ die Halbgruppe M erzeugt.

Wir schreiben: $[M_1, I_1]/[M_2, I_2]$ falls es ein Paar $[M_3, I_3]$ gibt mit

1) $[M_3, I_3]$ ist eine Halbgruppe mit Generatornamen
2) M_3 ist Unterhalbgruppe von M_2
3) Es gibt einen Halbgruppenepimorphisuus

$$h_1\colon M_3 \to M_1$$

4) $I_3 \subseteqq I_2$
5) Es gibt eine Bijektion $h_2\colon I_3 \to I_1$ mit

$$\beta_1 h_2(x) = h_1\beta_2(x) \quad (x \in I_3),$$

wobei β_1, β_2 zugehörige Namensfunktionen sind.

Zu einem Automaten a können wir in natürlicher Weise eine Halbgruppe mit Generatornamen in folgender Weise definieren.

Sei

$$M(a) = \{\delta_a^w \mid w \in I(a)^*\}$$

die Transitionshalbgruppe von a, dann ist $[M(a), I(a)]$ mit der Namensfunktion $\beta(x) = \delta_a^x$ ein derartiges Paar.

Korollar 3. *Ist* $h\colon a \Rrightarrow a'$ *eine Simulation und* a *vollreduziert, so gilt*

$$[M(a), I(a)]/[M(a'), I(a')] .$$

Wir vertiefen den Zusammenhang durch folgende Beobachtung.

Sei $[H, I]$ eine Halbgruppe mit Generatornamen und Namensfunktion β. Definiere $a[H, I]$ ausgabefrei durch

1) $I(a[H, I]) = I,\ S(a[H, I]) = H$
2) $\delta_{a[H,I]}(i, h) = h\beta(i) \quad (i \in I, h \in H)$.

Beobachtung.

1) $[H_1, I_1]/[H_2, I_2] \Rightarrow a[H_1, I_1]/a[H_2, I_2]$.

2) a/a' & a vollreduziert $\Rightarrow a[M(a), I(a)]/a[M(a'), I(a')]$.

3) a zyklisch $\Rightarrow a/a[M(a), I(a)]$.

Wir geben eine weitere Charakterisierung der Simulierbarkeit mit Hilfe der δ-Homomorphie.

Lemma 2.5. *Seien a und a' Automaten; dann sind folgende Aussagen äquivalent.*

(1) *Es gibt ein Paar (h_I, h_S) von injektiven Abbildungen $h_j: j(a) \to j(a')$ $(j = I, S)$ mit* $\delta_{a'}(h_I(w), h_S(s)) = \delta_{a'}(h_I(w'), h_S(s')) \Rightarrow \delta_a(w, s) = \delta_a(w', s')$ $(w, w' \in I(a)^*, s, s' \in S(a))$

(2) *Es gibt einen Automaten a'', einen δ-Monomorphismus $\varkappa: a'' \overset{(\delta)}{\to} a'$ und einen (I, S)-Epimorphismus $\varphi: a'' \to a$ mit φ_I bijektiv.*

Beweis. (2) $\Rightarrow$ (1). Definiere h_I und h_S durch $h_I = \varkappa_I \varphi_I^{-1}$ und $h_S = \varkappa_S \varrho$, wobei ϱ rechtsinvers zu φ_S ist; dann ergibt sich sofort die in (1) geforderte Eigenschaft.

(1) $\Rightarrow$ (2). Definiere a'' durch

1) $I(a'') = h_I(I(a))$,
$S(a'') = \{\delta_{a'}(h_I(w), h_S(s)) \mid s \in S(a), w \in I(a)^*\}$, $O(a'') = O(a)$;

2) $\delta_{a''}(h_I(x), s') = \delta_{a'}(h_I(x), s')$,
$\lambda_{a''}(h_I(x), s') = \lambda a(x, \delta_a(w, s'))$ mit $s' = \delta_{a'}(h_I(w), h_S(s))$
$(x \in I(a), s' \in S(a''))$.

Nach Voraussetzung ist a'' wohldefiniert, und es gilt $a'' \overset{(\delta)}{\leqq} a'$. Dann definiere $\varphi: a'' \to a$ durch

$$\varphi_I(x') = h_I^{-1}(x') \quad (x' \in I(a''))$$

und

$$\varphi_S(s') = \delta_a(w, s), \quad \text{falls } s' = \delta_{a'}(h_I(w), h_S(s)) \quad (s' \in S(a'')) .$$

Man erkennt sofort, daß φ wohldefiniert ist. — Nun folgt die Behauptung, indem man die Homomorphieeigenschaft nachrechnet.

Wir zeigen nun ein Zerlegungstheorem:

Theorem 2.2. *Zu jedem Automaten a existieren zyklische ausgabefreie Automaten $a_1, \ldots, a_k$ mit*

1) $a/(a_1 \oplus 1_{I(a)}) \otimes \cdots \otimes (a_k \oplus 1_{I(a)})$

2) $[M(a_i), I(a)]/[M(a), I(a)]$ *für alle* $1 \leqq i \leqq k$.

Beweis. Eine Teilmenge S' von $S(a)$ heißt Erzeugendensystem von a, falls es zu jedem $s \in S(a)$ ein $s' \in S'$ und $w \in I(a)^*$ mit $\delta_a(w, s') = s$ gibt.

Betrachte ein Erzeugendensystem $S' = \{s_1, \ldots, s_k\}$ von a. Definiere nun a_i $(1 \leqq i \leqq k)$ ausgabefrei durch

1) $I(a_i) = I(a)$, $S(a_i) = \{\delta_a(w, s_i) \mid w \in I(a)^*\}$,

2) $\delta_{a_i}(x, s) = \delta_a(x, s)$ $(s \in S(a_i), x \in I(a))$.

Da $a_i \overset{(\delta)}{\leqq} a$ für $1 \leqq i \leqq k$ gilt, folgt sofort

$$[M(a_i), I(a)]/[M(a), I(a)] .$$

Ferner sind nach Definition alle a_i zyklisch. Betrachte nun

$$b = (a_1 \oplus 1_{I(a)}) \otimes \cdots \otimes (a_k \oplus 1_{I(a)}) .$$

Sei

$$S'' = \{(\$, \ldots, \$, \underset{i}{s}, \$, \ldots, \$) \mid s \in S(a_i), 1 \leqq i \leqq k\},$$

dann gilt $\delta_b(I(b), S') \subseteqq S'$. Also definiert S'' einen Unterautomaten b' von b. Definiere nun h_1 durch:

$$h_1(\$, \ldots, \$, \underset{i}{s}, \$, \ldots, \$) = s \quad (s \in S(a_i), 1 \leqq i \leqq k).$$

Mit $h_2 = 1_{I(a)}$, folgt dann a/b nach Lemma 2.5.

3. Lineare Simulierbarkeit

Wir wenden die im Abschnitt 2 erzielten Resultate auf Realisierungsprobleme für Automaten an.

Definition 3.1 [6]. Sei R ein endlicher Ring mit Einselement. Ein Automat a ist *R-(links-)linear*, falls $j(a)$ unitärer *R-(links)-modul* für $j = I, O, S$ und δ_a und λ_a R-Modulhomomorphismen sind.

Wir studieren verschiedene Realisierungsbegriffe. Hierzu benötigen wir

Lemma 3.1. *Ist l R-linear, so ist* Red(l) *R-linear.*

Beweis. Bekanntermaßen ist $\mathrm{Red}_S(l)$ R-linear [6], also können wir o.B.d.A. annehmen, daß l reduziert ist. Betrachte nun

$$M = \{x \in I(l) \mid x \sim 0\}.$$

Sei nun $\delta_l(x, s) = h_l x + g_l s$ und $\lambda_l(x, s) = f_l x + j_l s$ $(x \in I(l), s \in S(l))$ die übliche Zerlegung von δ_l und λ_l in lineare Abbildungen. Offenbar gilt dann

$$x \sim x' \Leftrightarrow (x - x') \in \mathrm{Kern}\ h_l \cap \mathrm{Kern}\ f_l.$$

Nun gilt aber $M = \mathrm{Kern}\ h_l \cap \mathrm{Kern}\ f_l$. Hieraus folgt sofort die Behauptung.

Lemma 3.2. *Ist l vollreduziert und R-linear, so ist jeder Unterautomat von l vollreduziert.*

Beweis. Sei $a \subseteqq l$, dann ist a reduziert [5]. Sei nun $\delta_a(x, s) = \delta_a(x', s)$ und $\lambda_a(x,s) = \lambda_a(x', s)$ für $x, x' \in I(a)$ und alle $s \in S(a)$. Dann gilt (siehe Beweis von Lemma 3.1) $h_l(x) = h_l(x')$ und $f_l(x) = f_l(x')$. Hieraus folgt

$$\delta_l(x, s) = \delta_l(x', s) \quad \text{und} \quad \lambda_l(x, s) = \lambda_l(x', s)$$

für alle $s \in S(l)$. Also $x = x'$, da l vollreduziert.

Theorem 3.1. *Sei a ein Automat; dann sind folgende Aussagen äquivalent:*

1) *Es existiert a' und l R-linear mit $a \approx a'$ & $a' \leqq l$.*

2) *Es existiert l R-linear mit* Red(a) $\leq l$.

Beweis. Wir haben (1) ⇒ (2) zu zeigen. Sei also $a \approx a' \leq l$, l R-linear, gegeben. Betrachte die kanonische Reduktion $\varrho_v: l \to \mathrm{Red}(l)$. Dann gilt für $a'' = \varrho_v(a')$: $a' \approx a''$. Nach Lemma 3.2 ist a'' vollreduziert. Also gilt:

$$\mathrm{Red}(a) \triangleq \mathrm{Red}(a') \triangleq \mathrm{Red}(a'') \triangleq a''.$$

Dann folgt mit Lemma 3.1 die Behauptung.

Wir nennen a' eine *I-Erweiterung* von a, falls es einen I-Epimorphismus $\varphi: a' \to a$ gibt.

Lemma 3.3. *Sei a' eine I-Erweiterung von a, es möge ferner ein R-linearer Automat l existieren mit $a \leqq l$, dann existiert ein R-linearer Automat l' mit $a' \leq l'$.*

Den einfachen Beweis übergehen wir.

Beachtet man nun, daß $\mathrm{Red}_S(a)$ stets eine I-Erweiterung von $\mathrm{Red}(a)$ ist, so erhalten wir das

Korollar. *Sei a ein Automat; dann sind folgende Aussagen äquivalent:*

(1) *Es existiert a' und l R-linear mit*

$$a \sim a' \;\&\; a' \leq l\,.$$

(2) *Es existiert a' und l R-linear mit*

$$a \approx a' \;\&\; a' \leq l\,.$$

Beweis. Nach dem Satz von Gill [5], ist (1) äquivalent zu:

„Es existiert l R-linear mit $\mathrm{Red}_S(a) \leq l$."

Hieraus folgt mit Theorem 3.1 die Behauptung.

Theorem 3.2. *Sei a ein Automat; dann sind folgende Aussagen äquivalent:*

(1) *Es existieren a' und l R-linear mit $a \sim a' \;\&\; a' \overset{(\delta)}{\leq} l$.*

(2) *Es existieren a' und l R-linear mit $a \sim a' \;\&\; a' \overset{(\delta)}{\leq} l$.*

(3) *Es existiert l R-linear mit a/l.*

Beweis. Es genügt, (3) $\Rightarrow$ (1) zu zeigen. Wir können o.B.d.A. annehmen, daß l ausgabefrei und a vollreduziert ist. Dann existieren nach Lemma 2.4 injektive Abbildungen h_I und h_S mit

$$\delta_l(h_I(w), h_S(s)) = \delta_l(h_I(w'), h_S(s')) \Rightarrow \delta_a(w, s) = \delta_a(w', s')$$

für alle $w, w' \in I(a)^*$ und $s, s' \in S(a)$.

Unter Benutzung der Tatsache, daß $\mathrm{Red}_S(a)$ I-Erweiterung von $\mathrm{Red}(a)$ ist, folgt mit Lemma 2.5 die Behauptung.

Wir studieren ein Beispiel, das zeigt, daß die Realisierungsbegriffe aus Theorem 3.2 allgemeiner als die aus Theorem 3.1 sind. Dazu betrachten wir die Zustandsgraphen aus Abb. 2.

a_0 ist vollreduziert, aber es gibt keinen endlichen Ring R und R-linearen Automaten l mit $a_0 \overset{(\delta)}{\leq} l$ (vgl. [7] und [13]). Nun gilt aber $a_0 \sim a_1$.

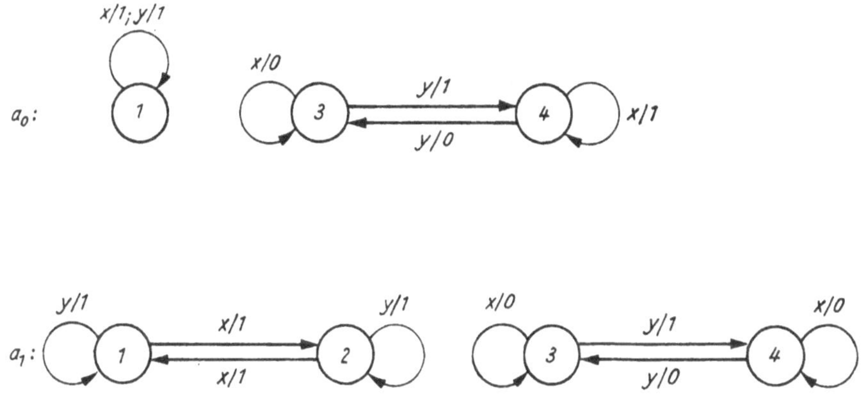

Abb. 2

Bestimmt man andererseits l ausgabefrei durch $I(l) = \mathbf{Z}_2$, $S(l) = \mathbf{Z}_2^2$ und

$$\delta_l\left(x, \begin{pmatrix} z_1 \\ z_2 \end{pmatrix}\right) = \begin{pmatrix} x + z_2 \\ x + z_1 \end{pmatrix} \quad (x, z_1, z_2 \in \mathbf{Z}_2)$$

so gilt: $a_1 \overset{(\delta)}{\triangleq} l$.

Wir untersuchen nun die Frage, wie sich die Automaten a charakterisieren lassen, zu denen es ein l R-linear mit a/l gibt. Hierzu verwenden wir die in Lemma 2.4 und 2.5 gegebene Charakterisierung.

Wir stellen zunächst einige einfache Beobachtungen zusammen.

Beobachtung 1. *Seien a_1 und a_2 Automaten, zu denen es l_1 und l_2 R-linear gibt mit a_1/l_1 und a_2/l_2, dann existieren l und l' R-linear mit*

$$a_1 \oplus a_2/l \qquad \text{und} \qquad a_1 \otimes a_2/l' ,$$

sofern die Operationen definiert sind.

Den einfachen Beweis übergehen wir und vermerken

Beobachtung 2. *Zu jedem Punktautomaten 1_X existiert ein l R-linear mit $1_X/l$.*

Ferner gilt

Beobachtung 3. *Ist a ein Automat, zu dem es l R-linear mit a/l gibt, so gibt es zu jeder I-Erweiterung a' von a ein l' R-linear mit a'/l'.*

Ausgangspunkt unserer Untersuchungen ist eine Charakterisierung der Transitionshalbgruppen linearer Automaten. Wir referieren kurz die Ergebnisse (vgl. [3], [4]). Dazu betrachten wir die allgemeine Transitionsformel [6]. Sei also l ein R-linearer Automat. Sei ferner $\delta_l(x, s) = h_l x + g_l s$ die kanonische Zerlegung, dann gilt für alle $w = x_0 \ldots x_{i-1}$ mit $x_\lambda \in I(l)$ für $0 \leqq \lambda \leqq i - 1$ und $s \in S(a)$

$$\begin{aligned} \delta_l(w, s) &= \sum_{\lambda=0}^{i-1} g_l^{i-\lambda-1} h_l x_\lambda + g_l^i s = \\ &= \delta_l(0, w) + g_l^i s . \end{aligned}$$

Sei nun H die zyklische Halbgruppe

$$H = \{g_l, g_l^2, \ldots\}$$

und

$$S = S(l)$$

die abelsche Gruppe der Zustände.

Wir wollen zeigen, daß $M\ (l)$ im wesentlichen ein semidirektes Produkt von H und S ist. Hierzu definieren wir: Seien H_1 und H_2 zwei Halbgruppen, $\mu\colon H_2 \to \operatorname{Endo}(H_1)$ ein Halbgruppenhomomorphismus in die Endomorphismenhalbgruppe von H_1. Dann ist $H_1 \underset{\mu}{\times} H_2$ definiert durch

$$(h_1, h_2) \cdot (h_1', h_2') = \big(h_1 \mu(h_2)\,(h_1'),\, h_2 h_2'\big)\ (h_1, h_1' \in H_1,\, h_2, h_2' \in H_2)\,.$$

$H_1 \underset{\mu}{\times} H_2$ ist wieder eine Halbgruppe, das *semidirekte Produkt* von H_1 und H_2 mit dem *Verbindungshomomorphismus* μ.

Bezeichnung. Ist H eine Halbgruppe, so ist H^{rev} definiert durch
$h_1 * h_2 = h_2 \cdot h_1 \quad (h_1, h_2 \in H)$.

Betrachten wir nun wieder l, sowie die Halbgruppen H und S. Definiert man nun

$$\mu\colon H \to \operatorname{Endo}(S)$$

durch

$$\mu(s, g_l^j) = g_l^j(s)$$

und

$$\varphi\colon M(l) \to S \underset{\mu}{\times} H$$

durch

$$\varphi(\delta_l^w) = \big(\delta_l(0, w), g_l^{L(w)}\big)$$
$$\big(w \in I(l)\big)^* ,$$

so ist φ ein Antihomomorphismus mit

$$\varphi(\delta_l^x) \in S \underset{\mu}{\times} \{g_l\} \text{ für } x \in I(l) .$$

Beachtet man nun noch die Eigenschaft

„Ist $s \neq 0$, $s \in S$, so gilt: Ordnung(s)/char(R)" ,

so erhalten wir das

Theorem 3.3 (vgl. [4]). *Sei a ein vollreduzierter Automat, zu dem es l R-linear mit a/l gibt, dann gibt es eine zyklische Halbgruppe H, eine abelsche Gruppe S, in der die Ordnung jedes von 0 verschiedenen Elements* char(R) *teilt, einen Verbindungshomomorphismus μ und einen Erzeuger $g \in H$ mit*

$$[M(a), I(a)]/[(S \underset{\mu}{\times} H)^{\mathrm{rev}}, S \times \{g\}] .$$

Hiervon wollen wir die Umkehrung zeigen. Sei also ein Automat a gegeben, sei ferner S eine abelsche Gruppe mit

(1) Für alle $s \in S$ mit $s \neq 0$ gilt: Ordnung(s)/char(R),

H eine zyklische Halbgruppe mit Generator g und μ ein Verbindungshomomorphismus, so daß

(2) $[M(a), I(a)]/[(S \underset{\mu}{\times} H)^{\mathrm{rev}}, S \times \{g\}]$

gilt.

Wir wollen zunächst die Aufgabenstellung reduzieren. Dazu betrachten wir den Zerlegungssatz Theorem 2.3, d. h., es existieren zyklische, ausgabefreie Automaten $a_1, \dots, a_k$ mit

(3) $a/(a_1 \oplus 1_{I(a)}) \otimes \cdots \otimes (a_k \oplus 1_{I(a)})$

und

(4) $[M(a_i), I(a)]/[M(a), I(a)]$ $(1 \leqq i \leqq k)$.

Nach den vorher angegebenen Beobachtungen, können wir also ohne Beschränkung der Allgemeinheit annehmen, daß a zyklisch ist. Nun gilt für zyklische Automaten a

(5) $a/a\, [M(a), I(a)]$.

Ferner gilt:

(6) $a[M(a), I(a)]/a[(S \underset{\mu}{\times} H)^{\mathrm{rev}}, S \times \{g\}]$.

Also genügt es zu zeigen, daß ein l R-linear existiert mit

$$a[(S \underset{\mu}{\times} H)^{\mathrm{rev}}, S \times \{g\}]/l .$$

Nach einem Satz von Stucky [13] gilt nun:

Seien R und R' zwei endliche Ringe mit char(R)/char(R'), dann existiert zu jedem l R-linear ein l' R'-linear mit l/l'.

Also können wir ohne Beschränkung der Allgemeinheit annehmen:

(7) $R = \mathsf{Z}_t$ mit $t = \mathrm{kgV}\{\mathrm{Ordnung}(s) \mid s \neq 0,\ s \in S\}$.

Nun gilt:

Lemma 3.4. *Unter den obigen Voraussetzungen existiert ein Z_t-linearer Automat l mit*

$$a[(S \underset{\mu}{\times} H)^{\mathrm{rev}}, S \times \{g\}]/l\,.$$

Beweis. Aufgrund der Wahl von t können wir S als Z_t-Modul auffassen. Sei nun $H = \{g, g^2, \ldots, g^m\}$ und $1 \leqq j(H) \leqq m$ mit $g^{m+1} = g^{j(H)}$.

Sei e_i der i-te Einheitsvektor in Z_t^m $(1 \leqq i \leqq m)$.

Definiere nun l ausgabefrei durch

1) $I(l) = S, \quad S(l) = S \times \mathsf{Z}_t^m$

2) $$\delta_l\left(s', \begin{pmatrix} s \\ e_i \end{pmatrix}\right) = \begin{cases} \begin{pmatrix} s'\mu(g)\,(s) \\ e_{i+1} \end{pmatrix} & \text{für } i < m\,, \\ \begin{pmatrix} s'\mu(g)\,(s) \\ e_{j(H)} \end{pmatrix} & \text{für } i = m \end{cases}$$

$(s, s' \in S,\ 1 \leqq i \leqq m)$. Definiere ferner $h_2\colon S \to S \times \{g\}$ bijektiv durch $h_2(s) = (s, g)$ und

$$h_1\colon \left\{\begin{pmatrix} s \\ e_i \end{pmatrix} \,\middle|\, s \in S\,, \quad 1 \leqq i \leqq m\right\} \to S \underset{\mu}{\times} H$$

durch

$$h_1\left(\begin{pmatrix} s \\ e_i \end{pmatrix}\right) = (s, g^i) \qquad (1 \leqq i \leqq m,\ s \in S)\,.$$

Wir zeigen

$$h_1\left(\delta_l\left(s', \begin{pmatrix} s \\ e_i \end{pmatrix}\right)\right) = \delta_{a[(S \underset{\mu}{\times} H)^{\mathrm{rev}},\, S \times \{g\}]}\left(h_2(s'), h_1\begin{pmatrix} s \\ e_i \end{pmatrix}\right)$$

für alle $s \in S$ und $1 \leqq i \leqq m$. Hieraus folgt dann die Behauptung.

1. Fall. $i < m$:

$$h_1\left(\delta_l\left(s', \begin{pmatrix} s \\ e_i \end{pmatrix}\right)\right) = h_1\begin{pmatrix} s'\mu(g)\,(s) \\ e_{i+1} \end{pmatrix} = \left(s'\mu(g)\,(s), g^{i+1}\right) = (s, g^i) \ast (s', g) =$$

$$= \delta_{a[(s \underset{\mu}{\times} H)^{\mathrm{rev}},\, S \times \{g\}]}\left((s', g), (s, g^i)\right)\,.$$

Da $(s, g^i) = h_1\begin{pmatrix} s \\ e_i \end{pmatrix}$ und $(s', g) = h_2(s')$ gilt, folgt in diesem Fall die Behauptung.

2. Fall. $i = m$:

$$h_1\left(\delta_l\left(s', \begin{pmatrix} s \\ e_m \end{pmatrix}\right)\right) = h_1\begin{pmatrix} s'\mu(g)\,(s) \\ e_{j(H)} \end{pmatrix} = \left(s'\mu(g)\,(s), g^{j(H)}\right) = \left(s'\mu(g)\,(s), g^{m+1}\right) =$$

$$= (s, g^m) \ast (s', g)\,.$$

Nun folgt wie im Fall 1 die Behauptung.

Wir fassen zusammen zum

Theorem 3.4. *Sei a ein Automat, R ein endlicher Ring mit Einselement.*

Es mögen eine abelsche Gruppe S und eine zyklische Halbgruppe H existieren, wobei gilt

(1) *Für alle* $s \neq 0$, $s \in S$ *gilt*: $\mathrm{Ordnung}(s)/\mathrm{char}(R)$

(2) *Es existiert ein Verbindungshomomorphismus μ und ein Generator g von H mit*

$$[M(a), I(a)]/[(S \underset{\mu}{\times} H)^{\mathrm{rev}}, S \times \{g\}] ,$$

dann gibt es einen R-linearen Automaten l mit a/l.

Wir studieren Beispiele.

Beispiel 1. Betrachte autonome, streng zusammenhängende Permutationsautomaten a. In diesem Fall gilt

$$M(a) \triangleq \mathsf{Z}_k \quad \text{mit } k = \#\,(S(a)) .$$

Betrachte nun Primzahl p mit $p \mid k$, $p^2 \nmid k$. Dann gilt

$$M(a) \triangleq \mathsf{Z}_p \times \mathsf{Z}_{k \cdot p^{-1}} .$$

In diesem Fall brauchen wir uns um den Erzeuger nicht weiter zu kümmern.

Die Anwendung der Konstruktion aus Theorem 3.4 liefert nun einen linearen Automaten l mit $\dim_{\mathsf{Z}_p} \cdot S(l) = 1 + p^{-1} \cdot k$.

Die einfachste Methode, autonome Automaten linear zu realisieren [13] liefert einen linearen Automaten l' mit $\dim_{\mathsf{Z}_p} (S(l')) = k$, d. h., das hier besprochene Verfahren ist in einigen Fällen wesentlich günstiger.

Beispiel 2. Betrachte Halbgruppe $H_0 = \{h_1, \dots, h_n\}$ mit $h_i h_j = h_i$ $(1 \leqq i, j \leqq n)$ und $a_{H_0} = a\,[H_0, H_0]$. Sei R Ring mit Einzelelement. Setze nun $S = R^n$ und $H = (g, g^0\}$ mit $g^2 = g$. Dann sei $\mu(g) \equiv 0$.

In $(S \underset{\mu}{\times} H)^{\mathrm{rev}}$ betrachte die Unterhalbgruppe

$$G_0 = \{(e_i, g) \mid 1 \leq i \leq n\} \cup \{0, g^0)\}) .$$

Dann rechnet man nach:

$$M(a_{H_0}) \triangleq G_0 .$$

Also sind diese Automaten über jedem Ring linear simulierbar.

Beispiel 3. Nach Theorem 3.4 muß notwendig gelten $M(a)/(S \underset{\mu}{\times} H)^{\mathrm{rev}}$ mit S abelsch und H zyklische Halbgruppe. Betrachten wir a_n mit $I(a_n) = \{x, y\}$, $S(a_n) = \{s_1, \dots, s_n\}$,

$$\delta_{a_n}(x, s_i) = s_{i+1} \qquad (1 \leqq i < n - 1) ,$$

$$\delta_{a_n}(x, s_{n-1}) = s_1 , \qquad \delta_{a_n}(x, s_n) = s_n ,$$

$$\delta_{a_n}(y, s_i) = s_i \qquad (1 \leqq i < n - 1) ,$$

$$\delta_{a_n}(y, s_{i-1}) = s_n ,$$

$$\delta_{a_n}(y, s_n) = s_{n-1} ,$$

so gilt $M(a_n) \triangleq \mathfrak{S}_n$ (symmetrische Gruppe der Ordnung n). Für $n \geqq 3$ ist aber die oben erwähnte Relation nicht möglich. Also gilt: a_n ist über keinem Ring linear simulierbar.

Literatur

[1] *Arbib, M. A.*, Algebraic Theory of Machines, Languages and Semi-Groups. Academic Press, New York 1968.
[2] *Arbib, M. A., H. P. Zeiger*, On the Relevance of Abstract Algebra to Control Theory. Automatica **5** (1969), 589–605.
[3] *Ecker, K.*, On the Semi-Group of a Linear Nonsingular Automaton. Math. Systems Theory **6** (1973), 353–358.
[4] *Ecker, K.*, Algebraische Eigenschaften linearer Automaten. Habilitationsschrift, Universität Bonn, 1973.
[5] *Gill, A.*, Linear Sequential Circuits: Analysis, Synthesis and Applications. McGraw-Hill, New York 1966.
[6] *Harrison, M. A.*, Lectures on Linear Sequential Machines. Academic Press, New York 1969.
[7] *Hartmanis, J., W. A. Davis*, Homomorphic Images of Linear Sequential Machines, J. Comp. Syst. Sci. **1** (1967).
[8] *Hartmanis, J., H. K.-G. Walter*, Group-theoretic Characterization of Linear Permutation Automata. J. Comp. Syst. Sci. **7** (1973), 168–188.
[9] *Herman, G.*, When is a Sequential Machine the Realization of another. Math. Systems Theory **5** (1971) 2, 115–127.
[10] *Hotz, G., H. K.-G. Walter*, Automatentheorie und formale Sprachen II. Endliche Automaten; BI, Hochschultaschenbücherverlag, Mannheim 1969.
[11] *Krohn, K., J. Rhodes*, Algebraic Theory of Machines. In: Proc. of the Symposion on Mathematical Theory of Automata, New York 1962; Polytechnic Press, Brooklyn 1963.
[12] *Reusch, B.*, Linear and Partial-linear Realizations of Automata. In: Proc. IFIP Congress, Edinburgh 1968, vol. 1; North-Holland, Amsterdam 1969; pp. 308–311.
[13] *Stucky, W.*, Linear realisierbare Automaten. Diss., Universität des Saarlandes, Saarbrücken, 1969.
[14] *Walter, H. K.-G.*, Synchronous Simulations of Sequential Machines. Berichte des mathematischen Instituts und Instituts für Angewandte Mathematik der Universität des Saarlandes, Saarbrücken, 1971, A 71/09.
[15] *Walter, H. K.-G.*, On Semi-Group Theoretical Techniques for the Linear-Realization Problem of Finite Automata. Berichte der Informatikforschungsgruppen des Fachbereichs Informatik der TH Darmstadt, Darmstadt, 1973, Ber. Nr. AFS 5.
[16] *Eichner, L.*, Homomorphe Darstellungen endlicher Automaten in linearen Automaten. EIK **9** (1973) 10, 587–613.

Kurzfassung

In dieser Arbeit wird eine Äquivalenztheorie für endliche Automaten entwickelt, die Simulationen als Vergleichsmethode verwendet. Hierdurch ergibt sich die Möglichkeit, halbgruppentheoretische Untersuchungen zur Lösung von Realisierungsproblemen heranzuziehen. In bezug auf lineare Realisierungen endlicher Automaten wird eine vollständige Charakterisierung linear realisierbarer Automaten hergeleitet. Dabei spielen im wesentlichen nur Struktureigenschaften der Transitionshalbgruppe eine Rolle.

Abstract

In this paper we give using simulations an equivalence theory of finite automata. This gives us the possibility to treat realization-problems with semigroup-theoretical methods. With respect to linear realizability of finite automata we derive a complete characterization of those automata which are linearly realizable. We show, that only structural properties of the transition monoid of an automaton determine wether or not the automaton is linearly realizable.

Резюме

В настоящей работе автор развивает теорию эквивалентности для конечных авроматов, которая в качестве метода сравнения использует моделирования. Таким образом получается возможность привлечения исследований в области теории полугрупп для решения проблем реализации. Относительно линейных реализаций конечных автоматов приводится полная характеристика реализуемых линейным образом автоматов. При этом в основном играют роль только свойства группы переходов состояний.

(Eingang: Erste Fassung am 20. 8. 1975,
ergänzte Fassung am 21. 6. 1976)

Anschrift des Verfassers:

Prof. Dr. H. K.-G. Walter
Fachbereich Informatik der
Technischen Hochschule Darmstadt
Forschungsgruppe Automatentheorie
und formale Sprachen
Magdalenenstr. 11
6100 Darmstadt
BRD

Elektronische Informationsverarbeitung und Kybernetik
EIK 13 (1977) 1/2, 55–60

Ein Beitrag zur Transformation ganzzahliger Optimierungsprobleme

Von *Gunter Bär*

1. Einleitung

In den letzten Jahren gab es zahlreiche Versuche, ein gegebenes ganzzahliges (lineares oder nichtlineares) Optimierungsproblem durch ein „äquivalentes" Optimierungsproblem zu ersetzen, das auf Grund der spezielleren Struktur der Nebenbedingungen leichter lösbar ist. Dabei seien zwei Optimierungsprobleme „äquivalent", falls ihre Zielfunktionen und ihre zulässigen Bereiche übereinstimmen. So überführen Bradley [5], Elmaghraby/Wig [6], Padberg [12], Glover/Woolsey [7] u. a. auf der Basis von Resultaten von Mathews [11] je zwei lineare (bzw. nichtlineare) Gleichungen eines ganzzahligen Optimierungsproblems in eine einzige lineare (bzw. nichtlineare) Gleichung mit den gleichen Variablen und den gleichen zulässigen Punkten. Durch mehrfache Anwendung dieser Beziehung entsteht aus einem gegebenen ganzzahligen Optimierungsproblem mit m Nebenbedingungen ein äquivalentes Optimierungsproblem mit nur einer Nebenbedingung. Von Nachteil ist bisher das starke Anwachsen der Größe der Koeffizienten in den neu entstehenden Nebenbedingungen, womit sich u. a. Glover/Woolsey [7] und Terzi [13] befaßten.

Vom Verfasser wurde in [3] eine nichtlineare Ungleichung mit 0–1-Variablen durch mehrere lineare Ungleichungen *mit den gleichen Variablen* und den gleichen zulässigen Punkten ersetzt. Damit läßt sich jedes nichtlineare 0–1-Optimierungsproblem in ein äquivalentes lineares 0–1-Optimierungsproblem umformen, allerdings mit sehr vielen Nebenbedingungen. Zum gleichen Resultat gelangen Granot/Hammer [8]; sie bezeichnen das äquivalente lineare 0–1-Optimierungsproblem als „verallgemeinertes Überdeckungsproblem". In [4] wird vom Verfasser eine Möglichkeit zur Reduzierung der Anzahl der Nebenbedingungen dargelegt.

Das in [3] bzw. [8] erhaltene Resultat — Ersetzung einer nichtlinearen Nebenbedingung mit 0–1-Variablen durch mehrere lineare Nebenbedingungen mit den gleichen Variablen — gilt i. allg. nicht für Nebenbedingungen mit beliebigen ganzzahligen Variablen. Nur dann, wenn der durch die gegebene nichtlineare Nebenbedingung (mit ganzzahligen Variablen) festgelegte zulässige Bereich ganzzahlig-konvex ist, existieren mit der Nebenbedingung äquivalente lineare Nebenbedingungen mit den gleichen Variablen.[1]) In der vorliegenden Arbeit soll gezeigt werden, daß eine beliebige Nebenbedingung (oder auch ein beliebiges System von Nebenbedingungen) eines ganzzahligen Optimierungsproblems mit beschränkten Variablen stets[2]) äquivalent durch spezielle quadratische Nebenbedingungen mit den gleichen Variablen ersetzt werden kann. Damit läßt sich jedes ganzzahlige Optimierungsproblem in ein äquivalentes quadratisches ganzzahliges Optimierungsproblem — ein sog. quadrati-

[1]) Siehe hierzu etwa Hadley [9].
[2]) Also auch, falls der zulässige Bereich nicht konvex ist.

sches Überdeckungsproblem — umformen. Der Beweis beruht im wesentlichen auf den in [1] dargelegten Beziehungen zwischen den Nebenbedingungen eines ganzzahligen Optimierungsproblems und den Ausdrücken des Aussagenkalküls. Gleichzeitig wird eine neue Möglichkeit zur Darstellung von Nebenbedingungen mit ganzzahligen Variablen in Form trennbarer Funktionen angegeben.

2. Transformation eines ganzzahligen Optimierungsproblems in ein quadratisches Überdeckungsproblem

Definition. a) Es seien $p_1, \dots, p_n$ logische Variablen. Dann sind p_i und $\bar{p}_i$ $(i = 1, 2, \dots, n)$ *Fundamentalalternativen (FA) der Variablen* $p_1, \dots, p_n$. Ist R FA der Variablen $p_1, \dots, p_n$ und tritt p_j $(j \in \{1, \dots, n\})$ weder negiert noch unnegiert in R auf, so sind auch $p_j \vee R$ sowie $\bar{p}_j \vee R$ *FA der Variablen* $p_1, \dots, p_n$.

b) Eine FA R heißt *Elementaralternative (EA) der Variablen* $p_1, \dots, p_n$, falls jede dieser Variablen genau einmal in R auftritt.

c) Ein Ausdruck $H(p_1, \dots, p_n)$ heißt *kanonische konjunktive Normalform der Variablen* $p_1, \dots, p_n$, wenn entweder $H \equiv W$ oder $H \equiv \bigwedge_{i=1}^{m} R_i$ ist, wobei R_i $(i = 1, \dots, m)$ EA der Variablen $p_1, \dots, p_n$ sind.

Der in [1] eingeführte und im folgenden benutzte Ableitungsbegriff „$\vdash H_1 \underset{H^*}{=} H_2$" besagt im wesentlichen, daß für jede sinnvolle Belegung der auftretenden Variablen der Wert von H_1 mit dem Wert von H_2 übereinstimmt und diese Äquivalenz rein formal aus einer gegebenen Axiomenmenge ableitbar ist. Sind H_1 und H_2 Nebenbedingungen eines ganzzahligen Optimierungsproblems, so ist $\vdash H_1 \underset{H^*}{=} H_2$ gleichbedeutend damit, daß H_1 und H_2 die gleichen zulässigen Punkte besitzen (also äquivalent sind).

Satz. *Es seien* z_j $(j = 1, \dots, l)$ *nichtnegative nach oben beschränkte ganzzahlige Variablen, die durch den Ansatz*

$$z_j = (1p_{j0} \vee 0\bar{p}_{j0}) + (2p_{j1} \vee 0\bar{p}_{j1}) + \cdots + (2^{k_j}p_{jk_j} \vee 0\bar{p}_{jk_j}) \tag{1}$$

mit $p \in \{F, W\}$ *in ein Tupel von logischen Variablen überführt werden können.*[1])

(Abkürzend sei mit A_j die rechte Seite von (1) bezeichnet.)

Behauptung. *Ist* $H \equiv p_{i_1} \vee \cdots \vee p_{i_r} \vee \bar{p}_{i_{r+1}} \vee \cdots \vee \bar{p}_{i_s}$ *eine EA der Booleschen Variablen* $p_{10}, \dots, p_{1k_1}, \dots, p_{l0}, \dots, p_{lk_l}$, *so gilt*

$$\vdash H \underset{H^*}{=} \sum_{j=1}^{l} (z_j - c_j)^2 \geqq 1\,,$$

wobei $\vdash c_j \underset{H^*}{=} A_j p_{i_1}/F \cdots p_{i_r}/F p_{i_{r+1}}/W \cdots p_{i_s}/W$ *für* $j = 1, \dots, l$ *sei.*

Zum Beweis wird der folgende Hilfssatz benötigt.

Hilfssatz. *Es sei für die ganzzahlige Variable z der Ansatz*

$$z = (1p_0 \vee 0\bar{p}_0) + (2p_1 \vee 0\bar{p}_1) + \cdots + (2^k p_k \vee 0\bar{p}_k) \tag{2}$$

gegeben und

$$\vdash c \underset{H^*}{=} A p_{i_0}/F \cdots p_{i_r}/F\, p_{i_{r+1}}/W \cdots p_{i_k}/W\,,$$

wobei $\{p_0, \dots, p_k\} = \{p_{i_0}, \dots, p_{i_k}\}$ *gelten soll.*

(A bezeichne die rechte Seite von (2).)

---

1) Zu den Beziehungen zwischen Zahlenvariablen und logischen Variablen siehe [1].

Behauptung. $\vdash (z = c) \underset{H^*}{=} \bar{p}_{i_0}\bar{p}_{i_1} \cdots \bar{p}_{i_r} p_{i_{r+1}} \cdots p_{i_k}$.

Beweis.

$$\vdash (z = c) \underset{H^*}{=} \sum_{\nu=0}^{k} (2^\nu p_\nu \vee 0\bar{p}_\nu)$$

$$= \left[\sum_{\nu=0}^{k} (2^\nu p_\nu \vee 0\bar{p}_\nu)\right] p_{i_0}/F \cdots p_{i_r}/F\; p_{i_{r+1}}/W \cdots p_{i_k}/W$$

$$\underset{H^*}{=} \sum_{\nu=0}^{k} (2^\nu p_\nu \vee 0\bar{p}_\nu) = \sum_{\nu=i_r+1}^{i_k} 2^\nu$$

$$\underset{H^*}{=} \sum_{\nu=i_0}^{i_r} 2^\nu \cdot (1p_\nu \vee 0\bar{p}_\nu) + \sum_{\nu=i_r+1}^{i_k} 2^\nu \cdot (1p_\nu \vee 0\bar{p}_\nu) = \sum_{\nu=i_r+1}^{i_k} 2^\nu$$

$$\underset{H^*}{=} \bar{p}_{i_0}\bar{p}_{i_1} \cdots \bar{p}_{i_r} p_{i_{r+1}} \cdots p_{i_k}.^{1)}\ \blacksquare$$

Zum Beweis des Satzes. Es sei $N_1 \equiv \sum_{j=1}^{l} (z_j - c_j)^2 < 1$; dann gilt

$$\vdash N_1 \underset{H^*}{=} \sum_{j=1}^{l} (z_j - c_j)^2 \leqq 0 \underset{H^*}{=} \sum_{j=1}^{l} (z_j - c_j)^2 = 0\ ^{1)}$$

$$\underset{H^*}{=} [(z_1 - c_1) = 0] \wedge [(z_2 - c_2) = 0] \wedge \cdots \wedge [(z_l - c_l) = 0]\ ^{1)}$$

$$\underset{H^*}{=} (z_1 = c_1) \wedge (z_2 = c_2) \wedge \cdots \wedge (z_l = c_l)\,.$$

Unter Benutzung des Hilfssatzes ergibt sich hieraus

$$\vdash N_1 \underset{H^*}{=} \bar{p}_{i_1} \cdots \bar{p}_{i_r} p_{i_{r+1}} \cdots p_{i_s}\,,$$

und nach Definition von H folgt nun $\vdash N_1 \underset{H^*}{=} \bar{H}$. Damit ergibt sich

$$\vdash H \underset{H^*}{=} \bar{N}_1 \underset{H^*}{=} \sum_{j=1}^{l} (z_j - c_j)^2 \geqq 1\,.\ \blacksquare$$

Die nach dem Satz entstehende und mit dem Ausdruck H äquivalente quadratische Nebenbedingung besitzt eine spezielle Gestalt, insbesondere stellt die linke Seite der Nebenbedingung stets eine trennbare[2]) Funktion dar.

Folgerung 1. *Ist $N(z_1, z_2, \ldots, z_l)$ eine beliebige nichtlineare Nebenbedingung eines ganzzahligen Optimierungsproblems (mit beschränkten Variablen), so existieren quadratische Nebenbedingungen $N_i(z_1, \ldots, z_l)$ mit $i = 1, \ldots, t$, mit der Eigenschaft*

$$\vdash N \underset{H^*}{=} \bigwedge_{i=1}^{t} N_i\,.$$

Beweis. Die Nebenbedingung $N(z_1, \ldots, z_l)$ kann unter Benutzung des Ansatzes (1) nach den Untersuchungen in [1] in einen äquivalenten Ausdruck $H(p_1, \ldots, p_s)$ des Aussagenkalküls überführt werden, und zu H gibt es eine äquivalente kanonische

---

[1]) Ggf. sind zu dem in [1] eingeführten Axiomensystem weitere Äquivalenzen als Axiome aufzunehmen, damit diese hier verwendeten Beziehungen rein formal abgeleitet werden können.

[2]) Die Funktion $f(z_1, \ldots, z_l)$ heißt *trennbar*, falls f als Summe von Funktionen darstellbar ist, die jeweils nur von einer Variablen abhängen (d. h. $f = f_1(z_1) + f_2(z_2) + \cdots + f_l(z_l)$).

konjunktive Normalform $K(p_1, \dots, p_s)$ der Gestalt $K \equiv \bigwedge_{i=1}^{t} H_i$, wobei H_i EA der Variablen $p_1, \dots, p_s$ sind. Jede EA H_i läßt sich nach obigem Satz in eine äquivalente quadratische Nebenbedingung $N_i(z_1, \dots, z_l)$ umformen, und hieraus ergibt sich die Behauptung. ∎

Beispiel. Gegeben sei die nichtlineare Nebenbedingung $N(z_1, z_2) \equiv z_1 \cdot z_2 \leqq 3$ mit $0 \leqq z_1, z_2 \leqq 3$ und ganzzahlig. Nach Ansatz (1) gilt etwa

$$z_1 = (1p_1 \vee 0\bar{p}_1) + (2p_2 \vee 0\bar{p}_2)$$

und

$$z_2 = (1p_3 \vee 0\bar{p}_3) + (2p_4 \vee 0\bar{p}_4)$$

mit $p \in \{F, W\}$. Dann liefern die Untersuchungen in [1], [2]

$$\vdash N \underset{H^*}{\equiv} (\bar{p}_1 \vee \bar{p}_2 \vee \bar{p}_3 \vee \bar{p}_4)(\bar{p}_1 \vee \bar{p}_2 \vee p_3 \vee \bar{p}_4)(p_1 \vee \bar{p}_2 \vee \bar{p}_3 \vee \bar{p}_4) \wedge$$
$$\wedge (p_1 \vee \bar{p}_2 \vee p_3 \vee \bar{p}_4),$$

und folglich ist $N(z_1, z_2)$ äquivalent mit dem folgenden Nebenbedingungssystem:

$$(z_1 - 3)^2 + (z_2 - 3)^2 \geqq 1,$$
$$(z_1 - 3)^2 + (z_2 - 2)^2 \geqq 1,$$
$$(z_1 - 2)^2 + (z_2 - 3)^2 \geqq 1,$$
$$(z_1 - 2)^2 + (z_2 - 2)^2 \geqq 1$$

und $0 \leqq z_1, z_2 \leqq 3$ und ganzzahlig.

Geometrisch besagt Folgerung 1, daß um jeden nichtzulässigen Gitterpunkt von $N(z_1, \dots, z_l)$, der in dem durch die Beschränkung der Variablen bestimmten Quader liegt, eine Hyperkugel mit dem Radius $r = 1$ gebildet wird. Der zulässige Bereich von N ergibt sich dann als Menge jener Gitterpunkte, die außerhalb beziehungsweise auf dem Rand aller erzeugten Hyperkugeln liegen.

Die Anzahl der zu einer gegebenen nichtlinearen Nebenbedingung N entstehenden speziellen quadratischen Nebenbedingungen ist gleich der Anzahl der nichtzulässigen Punkte von N und damit unter Umständen sehr groß. Allerdings läßt sich i. allg. bei mehreren nichtzulässigen Punkten von N auch die Anzahl der erforderlichen quadratischen Nebenbedingungen reduzieren, außerdem sind ggf. einige der entstehenden Nebenbedingungen linear. (Für das betrachtete Beispiel gilt etwa, daß N auch äquivalent ist mit $(z_1 - 3)^2 + (z_2 - 3)^2 \geqq 4$.) Genaueres hierzu soll in einer späteren Arbeit dargelegt werden.

Folgerung 2. *Ist* $\bigwedge_{i=1}^{k} N'_i$ *ein System von Nebenbedingungen eines ganzzahligen Optimierungsproblems (mit den beschränkten Variablen* $z_1, \dots, z_l$*), so existieren quadratische Nebenbedingungen* $N_i(z_1, \dots, z_l)$, $i = 1, \dots, t$, *mit* $\vdash \bigwedge_{i=1}^{k} N'_i \underset{H^*}{\equiv} \bigwedge_{i=1}^{t} N_i$.

Beweis. Jede gegebene Nebenbedingung N'_i wird — wie im Beweis von Folgerung 1 ausgeführt — in einen äquivalenten Ausdruck H'_i des Aussagenkalküls überführt. Dem Nebenbedingungssystem $\bigwedge_{i=1}^{k} N'_i$ entspricht dann der Ausdruck $H' \equiv \bigwedge_{i=1}^{k} H'_i$, der wiederum in eine kanonische konjunktive Normalform $K \equiv \bigwedge_{i=1}^{t} H_i$ umgeformt wird, wobei den EA H_i nach obigem Satz quadratische Nebenbedingungen der ursprünglich gegebenen ganzzahligen Variablen $z_1, \dots, z_l$ entsprechen. ∎

Unter der Voraussetzung $z_j \leqq 1$ für $j = 1, 2, \ldots, l$ liefert Folgerung 2 das von Granot/Hammer [8] als „verallgemeinertes Überdeckungsproblem" bezeichnete Resultat. Daher soll das in Folgerung 2 erhaltene Ergebnis auch so formuliert werden:

„Zu jedem nichtlinearen ganzzahligen Optimierungsproblem existiert stets ein äquivalentes quadratisches Überdeckungsproblem mit den gleichen ganzzahligen Variablen."

3. Abschließende Bemerkungen

Die hier betrachtete Transformation einer beliebigen nichtlinearen Nebenbedingung mit ganzzahligen Variablen in quadratische Nebenbedingungen mit den gleichen Variablen könnte vor allem für solche ganzzahligen Optimierungsprobleme genutzt werden, bei denen nur wenige nichtlineare Nebenbedingungen mit evtl. nur einigen Variablen vorhanden sind. Hier empfiehlt sich folgende Vorgehensweise:

- Umformung der vorhandenen nichtlinearen Nebenbedingungen N_i in einen Ausdruck K_i (bzw. eine konjunktive Normalform) des Aussagenkalküls. Ein Verfahren hierfür liefern die Untersuchungen in [1] und [2].
- Bildung der Konjunktion K aller konjunktiven Normalformen K_i, die äquivalent einer konjunktiven Normalform K' ist, und Aufstellung der quadratischen Nebenbedingungen N' auf der Grundlage der EA von K'.
- Lösung des neu erhaltenen Optimierungsproblems. Zur Lösung nichtlinearer ganzzahliger Optimierungsprobleme mit trennbaren Funktionen wurde von Krelle [10] ein Verfahren angegeben.

Es wurde folglich um jeden für die Gesamtheit aller nichtlinearen Nebenbedingungen nichtzulässigen Gitterpunkt genau eine Hyperkugel gebildet; d. h., redundante nichtlineare Nebenbedingungen werden bei der Bildung von K' automatisch erkannt und im folgenden berücksichtigt.

Zum Schluß sei darauf verwiesen, daß Folgerung 2 gleichzeitig eine neue Möglichkeit zur Darstellung von Nebenbedingungen mit ganzzahligen Variablen in Form trennbarer Funktionen liefert, die im Gegensatz zu Hadley [9] bzw. Krelle [10] zumindest keine zusätzlichen Variablen erfordert.

Literatur

[1] *Bär, G., H. Rohleder,* Über einen arithmetisch-aussagenlogischen Kalkül und seine Anwendung auf ganzzahlige Optimierungsprobleme. EIK **3** (1967) 3, 171–195.

[2] *Bär, G.,* Ein Verfahren zur Umformung einer linearen Nebenbedingung in eine äquivalente alternative Normalform. Z. Math. Logik Grundl. Math. **15** (1969), 163–180.

[3] *Bär, G.,* Zur linearen Darstellbarkeit von Ausdrücken des Aussagenkalküls. EIK 8 (1972) 6/7, 353–378.

[4] *Bär, G.,* Zur Linearisierung beliebiger 0–1-Optimierungsprobleme. Math. Operationsforsch. u. Statistik 7 (1976), 181–195.

[5] *Bradley, G. H.,* Transformation of integer programs to knapsack problems. Discr. Math. **1** (1971), 29–45.

[6] *Elmaghraby, S. E., M. K. Wig,* On the treatment of stock cutting problems as diophantine programs. Op. Res. Report No. 61, North Carolina State University at Raleigh, May 1970.

[7] *Glover, F., R. E. D. Woolsey,* Aggregating diophantine equations. Operations Res. **16** (1972), 1–10.

[8] *Granot, F., P. L. Hammer,* On the use of Boolean functions in 0–1-programming. Technion, Mimeograph Series on Operations Research, Statistics and Economics, No. 70, Sept. 1970.

[9] *Hadley, G.,* Nichtlineare und dynamische Programmierung. Verlag die Wirtschaft, Berlin 1969 (Übers. a. d. Englischen).

[10] *Krelle, W.*, Ganzzahlige nichtlineare Programmierung für trennbare Funktionen. In: Operations Research Verfahren, Bd. 5 (Herausg.: *R. Henn*); 1968; S. 220–236.
[11] *Mathews, G. B.*, On the partition of numbers. Proc. London math. Soc. **28** (1897), 486–490.
[12] *Padberg, M.*, Equivalent knapsack-type formulations of bounded integer linear programs: an alternative approach. Naval Res. Log. Quart. **19** (1972), 699–708.
[13] *Терзи Д. Г.*, Об агрегации уравнений. Прикладная математика и программирование, вып. 11; Москва 1974; 95–98.

Kurzfassung

Dargelegt wird, daß eine beliebige Nebenbedingung eines ganzzahligen Optimierungsproblems mit beschränkten Variablen stets äquivalent durch spezielle quadratische Nebenbedingungen mit den gleichen Variablen ersetzt werden kann. Folglich ist jedes ganzzahlige Optimierungsproblem äquivalent einem quadratischen ganzzahligen Optimierungsproblem. Dabei seien zwei Probleme äquivalent, falls ihre Zielfunktionen und ihre zulässigen Bereiche übereinstimmen.

Abstract

It is shown that any restriction of an integer programming problem with bounded variables always equivalent can be transformed into special quadratic restrictions with the same variables. Consequently every integer programming problem is equivalent to a quadratic integer programming problem. Two problems are said to be equivalent, if their target functions and their admissible ranges correspond.

Резюме

Излагается, что любое ограничение проблемы целочисленного программирования с ограниченными переменными всегда эквивалентно заменимо квадратными ограничениями специального вида, которые содержат те же самые переменные. Следовательно всякая проблема целочисленного программирования сводится к некоторой эквивалентной проблеме квадратного целочисленного программирования. Эквивалентными называются две проблемы, если их целевые функции и их допустимые области совпадают.

(Eingang: erste Fassung am 27. 3. 1975,
vorliegende Fassung am 10. 5. 1976)

Anschrift des Verfassers:

Dr. G. Bär
Sektion Mathematik der Karl-Marx-Universität
701 Leipzig
Karl-Marx-Platz 10/11
DDR

Elektronische Informationsverarbeitung und Kybernetik
EIK 13 (1977) 1/2, 61–77

Über den Einsatz mehrstufiger Mehrschrittalgorithmen zur Lösung von Adaptionsproblemen

Von *Klaus Fritzsch*[1])

0. Problemstellung

Iterationsverfahren finden bei der Lösung von Optimierungsaufgaben der verschiedensten Art Verwendung. Sie sind in der Regel als Einschrittverfahren konzipiert. Zur Berechnung des Schrittes von $\boldsymbol{w}[n-1]$ nach $\boldsymbol{w}[n]$ wird von den $\boldsymbol{w}[n-i]$ explizit nur $\boldsymbol{w}[n-1]$ benutzt.

Bei der numerischen Lösung von Differentialgleichungen werden mit Erfolg Mehrschrittverfahren eingesetzt, in denen zur Berechnung von $\boldsymbol{w}[n]$ die Zustände $\boldsymbol{w}[n-i]$ aus mehreren zurückliegenden Takten verwendet werden. Die Frage nach den Einsatzmöglichkeiten von Mehrschrittverfahren bei der Optimierung verdient daher eine gewisse Aufmerksamkeit.

Bei Adaptionsaufgaben, die in mathematischer Hinsicht eine spezielle Klasse von Optimierungsproblemen darstellen, läßt sich der Einsatz von Mehrschrittalgorithmen auch lerntheoretisch motivieren. Ein adaptives System wird unter gewissen Umständen sein Ziel schneller erreichen, wenn es zur Ermittlung des Zustandes $\boldsymbol{w}[n]$ nicht nur den Zustand $\boldsymbol{w}[n-1]$, sondern auch weiter zurückliegende Zustände heranzieht. Es nutzt damit die bereits vorhandenen Erfahrungen, soweit sie in den $\boldsymbol{w}[n-i]$ manifest sind.

Das Ziel der vorliegenden Arbeit ist es zu untersuchen, wie Adaptionsprozesse durch Verwendung von Mehrschrittprinzipien beschleunigt werden können. Zur Illustration werden auch Optimierungsprobleme in die Betrachtung einbezogen.

1. Adaptive Einschrittalgorithmen

Zahlreiche Aufgabenstellungen aus der Prozeßsteuerung und der Objekterkennung lassen sich wie folgt formalisieren. Es ist ein unimodales Gütefunktional

$$J(\boldsymbol{w}) = E_X[Q(\boldsymbol{x}, \boldsymbol{w})] \tag{1.1}$$

unter Beachtung von Nebenbedingungen

$$G_i(\boldsymbol{w}) = E_X[g_i(\boldsymbol{x}, \boldsymbol{w})]\,, \qquad i = 1, \ldots, i_0 \tag{1.2}$$

die die Unimodalität nicht beeinträchtigen, zu einem Minimum zu machen [1]. $E_X[\ldots]$ bezeichnet die Mittelung über die Zustände eines Prozesses oder die Objekte einer Menge. Das Optimierungsproblem läßt sich herkömmlich lösen, wenn die Mittelung über X vorab ausgeführt werden kann. Ist dies nicht möglich, muß die Lösung durch Adaption erfolgen. Im weiteren wird die herkömmliche Optimierung einfach als Optimierung und die adaptive Optimierung als Adaption bezeichnet.

---

[1]) Zentralinstitut für Kybernetik und Informationsprozesse der AdW.

Zur iterativen Lösung des Problems (1.1), (1.2) eignen sich die sog. Abstiegsverfahren. Sie haben die Gestalt

$$\boldsymbol{w}[n] = \boldsymbol{w}[n-1] + \gamma[n] \cdot \mathbf{s} . \tag{1.3}$$

In Optimierungsproblemen kann das Funktional (1.1) als Funktion von w aufgefaßt werden, demzufolge hängt $\mathbf{s}$ in (1.3) nur von $\boldsymbol{w}[n-i]$ mit $i = 1, \ldots, i_0$; $i_0 < n$ ab. In Adaptionsproblemen hingegen ist s eine Funktion der $\boldsymbol{w}[n-i]$ und $\boldsymbol{x}[n]$. Um Konvergenz zu gewährleisten, sind der Schrittweitenfolge $\gamma[n]$ gewisse einschränkende Bedingungen auferlegt.

Die Maßzahl

$$b = \frac{|\boldsymbol{s}^T \cdot \nabla_{\boldsymbol{w}} J(\boldsymbol{w})|}{|\boldsymbol{s}| \cdot |\nabla_{\boldsymbol{w}} J(\boldsymbol{w})|} \tag{1.4}$$

bzw.

$$b = \frac{|\boldsymbol{s}^T \cdot E_X[\nabla_{\boldsymbol{w}} Q(\boldsymbol{x}, \boldsymbol{w})]|}{|\boldsymbol{s}| \cdot |E_X[\nabla_{\boldsymbol{w}} Q(\boldsymbol{x}, \boldsymbol{w})]|} \tag{1.5}$$

dient als quantitative Angabe der Brauchbarkeit einer Abstiegsrichtung [2], [3]. Für optimal brauchbare Richtungen gilt $b = 1$. Eine Richtung ist gut brauchbar, falls die Aussage

$$b \ll 1 \tag{1.6}$$

falsch ist.

Der allgemeine Ansatz (1.3) wurde gewählt, um die Abstiegsverfahren formal einheitlich behandeln zu können. Diese lassen sich im wesentlichen in zwei Klassen, die Gradienten- und die Suchverfahren, einteilen. In der einfachsten Variante präsentieren sie sich als Einstufenverfahren, die durch eine Verstellregel, die Gleichung (1.3), vollständig charakterisiert sind.[1]) Die Schrittweite γ ist höchstens vom Takt abhängig. Die Bedingungen für die Konvergenz der einstufigen Verfahren sind gut bekannt. Daraus allein lassen sich keine effektiven Verstellregeln für $\gamma\mathbf{s}$ ableiten, so daß die Konvergenz, die in der Regel unbefriedigend ist, nicht in erforderlichem Maße beschleunigt werden kann.

Eine Verbesserung läßt sich nach Albert und Gardner [4] dadurch erzielen, daß ein allgemeinerer Ansatz für $\gamma\mathbf{s}$ gemacht wird, indem der Skalar γ durch eine Matrix Γ ersetzt wird. Γ ist i. allg. eine Funktion der $\boldsymbol{w}[k]$ und $\boldsymbol{x}[k]$. Aus (1.3) wird nun

$$\boldsymbol{w}[n] = \boldsymbol{w}[n-1] + \Gamma \cdot \mathbf{s} . \tag{1.7}$$

Mit Hilfe eines zweiten Gütekriteriums, das (1.1) untergeordnet ist, werden die gewonnenen Freiheitsgrade genutzt, um die Konvergenz zu beschleunigen. Das zweite Kriterium hat die Gestalt [4]

$$E_p[(\boldsymbol{w}[n] - \boldsymbol{w}^*)^T (\boldsymbol{w}[n] - \boldsymbol{w}^*)] \to \min . \tag{1.8}$$

$E_p[\ldots]$ bezeichnet die Mittelung über alle gemäß (1.7) möglichen Iterationsprozesse.

Daraus leiten sich Verstellregeln für die Matrix Γ ab. Sie sollen Verstellregeln zweiter Stufe heißen. Ein zweistufiges Abstiegsverfahren setzt sich aus je einer Verstellregel erster und zweiter Stufe zusammen. Die Hierarchie der beiden Regeln entspricht der der Gütekriterien.

Die Optimierung einschrittiger Iterationsalgorithmen wird ausführlich in der Monographie von Albert und Gardner behandelt. Diese Autoren verwenden ein Regressionsmodell, dem implizit die Annahme eines quadratischen Gütekriteriums zugrundeliegt.

$$J(\boldsymbol{w}) = E_x[(f(\boldsymbol{x}) - \boldsymbol{w}^T \cdot \varphi(\boldsymbol{x}))^2] \to \min . \tag{1.9}$$

[1]) Die Einstellung von Lagrangeschen Multiplikatoren für den Anschluß der Nebenbedingungen wird nicht als selbständige Verstellung betrachtet.

Der Abstieg zum Minimum verläuft bei den entsprechenden einstufigen Verfahren längs der jeweiligen Richtung des Antigradienten.

Bei den zweistufigen Verfahren wird Optimalität gemäß (1.8) gefordert, und es ergibt sich

$$\boldsymbol{w}[n] = \boldsymbol{w}[n-1] + \Gamma[n]\left(f(\boldsymbol{x}[n]) - \boldsymbol{w}[n-1]\,\varphi^T(\boldsymbol{x}[n])\right)\varphi(\boldsymbol{x}[n]) \tag{1.10}$$

mit

$$\Gamma[n] = \Gamma[n-1] - \frac{(\Gamma[n-1]\cdot\boldsymbol{\varphi})\,(\Gamma[n-1]\cdot\boldsymbol{\varphi})^T}{1+\boldsymbol{\varphi}^T\Gamma[n-1]\,\boldsymbol{\varphi}}\,. \tag{1.11}$$

Weitere Arbeiten zur Optimalität der Algorithmen stammen u. a. von Zypkin [5], Kesten [6] sowie Nevelson und Chasminski [7]. Zypkin [8] untersuchte den Zusammenhang zwischen der Art des Gütekriteriums erster Stufe und den Möglichkeiten der Optimierung der Algorithmen durch eine zweite Stufe. Es zeigte sich, daß sich die Optimierung in der zweiten Stufe streng nur bei quadratischen Kriterien erster Stufe durchführen läßt und eine weitere Verbesserung in diesem Falle unmöglich ist.

Höge [9] hat mit Hilfe von Rechnersimulation festgestellt, daß die Einführung neuer Freiheitsgrade in das inhomogene Glied $\gamma\boldsymbol{s}$ nur dann zweckmäßig ist, wenn genügend A-priori-Information zur Einstellung geeigneter Anfangswerte verfügbar ist.

Ein andersgearteter Zugang wird in den Arbeiten von Kesten [6] und Saridis [10] gewählt. Die Schrittweitensteuerung geschieht vermittels eines Rückkopplungsprinzips, das am Beispiel des Gradientenverfahrens erläutert sei. Wenn eine Komponente des Gradienten an aufeinanderfolgenden Takten das Vorzeichen beibehält, so ist daraus zu schließen, daß das Minimum in dieser Richtung noch relativ weit entfernt ist und die Schrittweite sinnvollerweise beibehalten [6] bzw. erhöht [10] werden sollte. Ändert sich das Vorzeichen jedoch von Takt zu Takt, so ist entweder das Minimum erreicht — die Vorzeichenwechsel beruhen auf Störungen oder kleinen Schwingungen — oder $\boldsymbol{w}[k]$ schwingt mit relativ hoher Amplitude. In diesen beiden Fällen ist es angezeigt, die Schrittweite zu vermindern.

Saridis hat die folgende Schrittweitensteuerung vorgeschlagen. $\Gamma[n]$ wird in der Form

$$\Gamma[n] = \gamma\cdot P[n] \tag{1.12}$$

angesetzt. γ ist konstant. $P[n]$ ist eine Diagonalmatrix mit den Gliedern $p_i[n]$, die der Verstellregel

$$p_i[n] = (1-\alpha)\,p_i[n-1] + \alpha\lambda_i[n]\,, \tag{1.13}$$

$$\lambda_i[n] = \begin{cases} 0\,, & \text{falls } \nabla_i[n]\cdot\nabla_i[n-1] \leqq 0 \wedge m \neq 0\,,\\[1ex] \dfrac{1}{m}\,, & \text{falls } \nabla_i[n]\cdot\nabla_i[n-1] > 0 \wedge m \neq 0\,,\\[1ex] \dfrac{1}{N}\,, & \text{falls } m = 0 \end{cases} \tag{1.14}$$

genügen. Dabei ist m die Zahl der Komponenten des Gradienten, die das Vorzeichen nicht geändert haben und $\nabla_i[n]$ die i-te Komponente des Gradienten zum Taktzeitpunkt $t = n$.

Analog lassen sich die auf stochastischen Suchmethoden beruhenden Einschrittalgorithmen zweistufig auslegen. Rastrigin [11] hat ein zweistufiges Verfahren mit einem steuerbaren Zufallsgenerator beschrieben. Die Wahrscheinlichkeit der Auswahl der i-ten Richtung wird gemäß

$$W_i[n] = W_i[n-1]\cdot(1-\alpha) + \alpha\tilde{\lambda}_i \tag{1.15}$$

bestimmt. Der Wert von $\tilde{\lambda}_i$ hängt von der Güteänderung beim vorangegangenen Schritt ab und zwar in der Weise, daß sich $W_i[n]$ für aussichtsreiche Richtungen erhöht und für ungünstige vermindert, wobei die Nebenbedingung

$$\sum_i W_i[n] = 1 \tag{1.16}$$

einzuhalten ist. Die Gleichungen (1.13) und (1.15) sind formal gleich. Sie unterscheiden sich lediglich in der Interpretation.

Das Rückkopplungsprinzip ist auch bei stochastischen Methoden zur Schrittweitensteuerung einsetzbar. Mit seiner Hilfe gelingt es, die Schwankungen nach Erreichen des Optimums einzuschränken. Die Anwendung der zweistufigen Einschrittverfahren wird gelegentlich erschwert oder verhindert. Optimale Matrizen $\Gamma[n]$ lassen sich exakt bei quadratischen Kriterien erster Stufe und sonst nur mehr oder weniger gut genähert finden. Für die Berechnung ist ein ziemlich hoher Aufwand zu treiben. Im Fall schlecht konditionierter Probleme ist zusätzlich eine Regularisierung auszuführen [5].

Obwohl die Verbesserungen, die die Schrittweitensteuerung durch Rückkopplung bringt, außer Zweifel stehen, darf ein Nachteil nicht unerwähnt bleiben. Jede Komponente des inhomogenen Gliedes wird getrennt verstellt und somit hängt die Wirksamkeit der Verstellregel zweiter Stufe von der Wahl des Koordinatensystems ab. Schlecht konditionierte Probleme mit Schluchten, die nicht parallel zu einer Koordinatenrichtung verlaufen, lassen sich damit nur bedingt optimieren. Die Empfindlichkeit des Algorithmus gegenüber Störungen steigt mit abnehmender Schrittweite. An den Knick- und Umkehrstellen der Trajektorie vergehen in der Regel einige Takte, bis eine günstige Richtung gefunden ist.

2. Einstufige Mehrschrittalgorithmen

Die Einführung der Mehrstufigkeit durch Vorgabe einer Hierarchie von Gütekriterien ist eine Erweiterungsmöglichkeit, eine andere die des Übergangs von Einschritt- auf Mehrschrittalgorithmen. Hier wird die Variante, daß weitere Glieder $\boldsymbol{w}[n-i]$ aufgenommen werden, untersucht:

$$\boldsymbol{w}[n] = \sum_{i=1}^{k} a_i \boldsymbol{w}[n-i] + \gamma \boldsymbol{s} \,. \tag{2.1}$$

Die inhomogene Erweiterung, die darin bestünde, mehrere Glieder $\gamma \boldsymbol{s}$ aus verschiedenen Takten aufzunehmen, wird hier nicht betrachtet.

Sind die Koeffizienten a_i während des Iterationsprozesses konstant und wird auch der inhomogene Term $\gamma \boldsymbol{s}$ nicht durch eine Verstellregel zweiter Stufe gesteuert, so liegt ein einstufiger Mehrschrittalgorithmus vor.

Mit dem folgenden Exkurs über einstufige Mehrschrittalgorithmen werden die Ausführungen zu den mehrstufigen Mehrschrittalgorithmen vorbereitet. Jene haben die allgemeine Gestalt von (2.1). Je nach der speziellen Wahl der a_i hat ein solcher Algorithmus die Fähigkeit, agil oder träge zu arbeiten, über Nebenminima hinwegzulaufen o. ä. Die Theorie der Mehrschrittalgorithmen hängt eng mit der Theorie der Differenzengleichungen zusammen. Der homogene Teil von (2.1) ist eine homogene lineare Differenzengleichung k-ter Ordnung:

$$\boldsymbol{w}[n] - \sum_{i=1}^{k} a_i \boldsymbol{w}[n-i] = 0 \,. \tag{2.2}$$

Es gilt die folgende Stabilitätsbedingung [12]: Das zu (2.2) gehörige charakteristische Polynom

$$\psi(\mu) = -\mu^k + a_1 \mu^{k-1} + \cdots + a_k \tag{2.3}$$

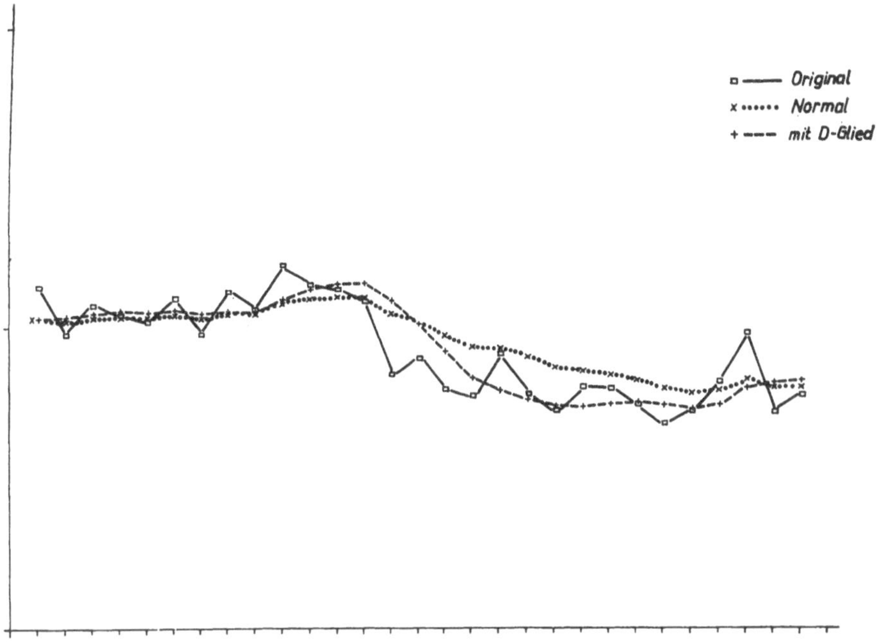

Abb. 1. Beispiel eines agilen Adaptionsalgorithmus
——— Verlauf von $x[n]$
Verlauf von $c[n]$ für Einschrittalgorithmus
– – – – Verlauf von $c[n]$ für Zweischrittalgorithmus mit positivem r_1

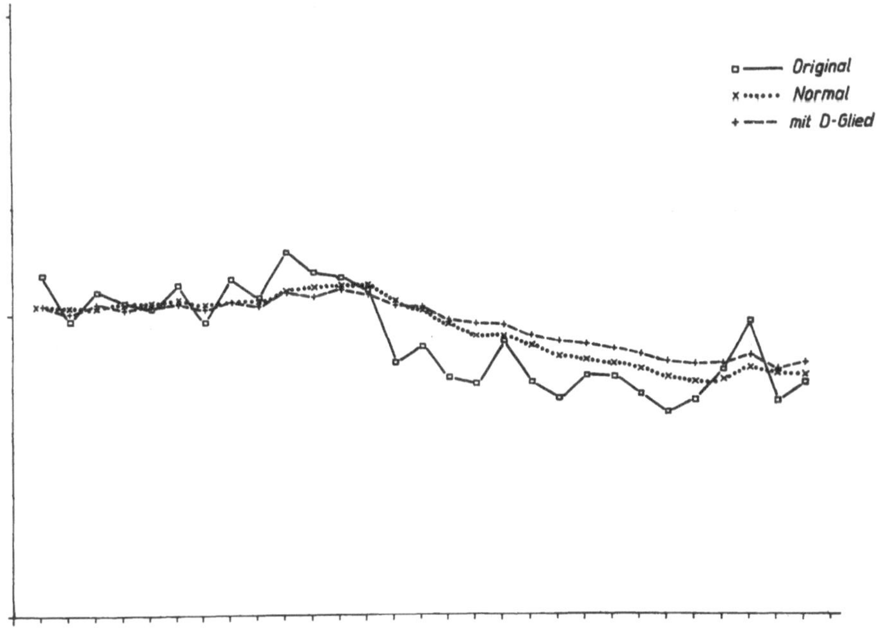

Abb. 2. Beispiel eines trägen Adaptionsalgorithmus
——— Verlauf von $x[n]$
Verlauf von $c[n]$ für Einschrittalgorithmus
– – – – Verlauf von $c[n]$ für Zweischrittalgorithmus mit negativem r_1

hat nur Wurzeln innerhalb oder auf dem Rande des Einheitskreises. Aus $\psi(\lambda) = 0$ und $|\lambda| = 1$ folgt, daß λ einfache Nullstelle von $\psi(\mu)$ ist.

Die Bedingung dafür, daß $\lambda = 1$ Nullstelle ist, lautet

$$\sum_{i=1}^{k} a_i = 1 . \tag{2.4}$$

Unter dieser Voraussetzung kann (2.1) umgeformt werden in

$$\begin{aligned} \boldsymbol{w}[n] = \boldsymbol{w}[n-1] &+ r_1(\boldsymbol{w}[n-1] - \boldsymbol{w}[n-2]) \\ &+ r_2\{(\boldsymbol{w}[n-1] - \boldsymbol{w}[n-2]) - (\boldsymbol{w}[n-2] - w[n-3])\} \\ &+ \cdots \\ &+ r_{k-1}\{\cdots\} + \gamma \boldsymbol{s} . \end{aligned} \tag{2.5}$$

Gleichung (2.5) läßt sich gut anschaulich interpretieren. Das Trendverhalten hängt hauptsächlich von r_1 ab. Positives r_1 bringt einen Zuschlag auf $\boldsymbol{w}[n-1]$ ein, der die Tendenz von $\boldsymbol{w}[n-2]$ nach $\boldsymbol{w}[n-1]$ weiterführt. Negatives r_1 bremst den Trend oder kehrt ihn um. In Zweischrittalgorithmen gibt es nur einen Freiheitsgrad, d. h.

$$r_1 \neq 0 \quad \text{und} \quad r_i = 0 \text{ für } i \geqq 2 . \tag{2.6}$$

Die Stabilitätsbedingung fordert

$$|r_1| < 1 . \tag{2.7}$$

Abb. 1 und 2 zeigen je eine Realisierung eines Zweischrittalgorithmus mit positivem bzw. negativem r_1. Das agile bzw. träge Verhalten ist gut zu erkennen.

Einstufige Zweischrittalgorithmen mit konstantem $r_1 > 0$ werden u. a. von Judin [13] zur Optimierung beim Vorhandensein von Nebenminima vorgeschlagen.

Der Koeffizient r_2 bestimmt, wie der „Trend des Trends" eingeht. Im Grenzfall $k = n$ geht der Mehrschrittalgorithmus in einen Optimierungsalgorithmus über. Sowohl aus theoretischen wie aus praktischen Gründen wird man sich jedoch auf wenige Schritte beschränken.

Formal und inhaltlich sind die Mehrschrittalgorithmen auch mit den autoregressiven Prozessen verwandt. In einem autoregressiven Prozeß ist jedoch das inhomogene Glied in der Regel unabhängig von den homogenen Termen. Trotz dieses und anderer Unterschiede ist eine wechselseitige Befruchtung der Theorie denkbar.

Der entscheidende inhaltliche Unterschied sowohl zu den Differenzengleichungen als auch zu den autoregressiven Prozessen ergibt sich aus dem angestrebten adaptiven Verhalten der Mehrschrittalgorithmen. Dieses Ziel wird dadurch erreicht, daß die r_i als veränderlich angenommen werden. Damit ist der Übergang zu den mehrstufigen Mehrschrittalgorithmen vollzogen.

Aus der Stabilitätstheorie der einstufigen Algorithmen ergeben sich gewisse Beschränkungen innerhalb der Klasse der mehrstufigen Mehrschrittalgorithmen. Im Falle der zweistufigen Algorithmen lassen sie sich mit relativ einfachen Mitteln formulieren.

3. Zweistufige Mehrschrittalgorithmen

Der methodische Ansatz bei der Erarbeitung der Verstellregel zweiter Stufe ist kybernetisch orientiert. Er benutzt das von Kesten [6] eingeführte Rückkopplungsprinzip. Die konkrete Gestalt der Algorithmen hängt von der Problemstellung ab. Die Adaption unter nichtstationären Bedingungen und die Optimierung bzw. Adaption bei schlecht konditionierten Problemen sind wichtige Anwendungsgebiete der Mehrschrittalgorithmen.

Hier soll das Problem der schlechten Kondition im Vordergrund stehen. Grundsätzlich ist dafür der folgende zweistufige Zweischrittalgorithmus geeignet:

$$\boldsymbol{w}[n] = \boldsymbol{w}[n-1] + r[n] \cdot (\boldsymbol{w}[n-1] - \boldsymbol{w}[n-2]) + \gamma \cdot \mathbf{s} , \tag{3.1}$$

$$r[n] = r[n-1]\,(1-\alpha_1) + \alpha_2 \cdot \operatorname{sign}\,(\Delta Q[n-1]) + \alpha_3 . \tag{3.2}$$

Dabei gilt

$$\Delta Q[n-1] = Q(\boldsymbol{x}[n-1], \boldsymbol{w}[n-2]) - Q(\boldsymbol{x}[n], \boldsymbol{w}[n-1])$$

bzw. (3.3)

$$\Delta Q[n-1] = J(\boldsymbol{w}[n-1]) - J(\boldsymbol{w}[n-2]) .$$

Die Analogie zur Schrittweitensteuerung durch Rückkopplung ist in (3.2) erkennbar. Die alte Adaptionsrichtung ist durch die Differenz $\boldsymbol{w}[n-1] - \boldsymbol{w}[n-2]$ gegeben. Deren Beitrag zur Verstellung von $\boldsymbol{w}[n-1]$ auf $\boldsymbol{w}[n]$ wird gewichtet. Erweist sich die eingeschlagene Abstiegsrichtung von einem gewissen Zeitpunkt an nicht mehr als günstig ($\Delta Q[n-1] < 0$), muß ein Bremsmechanismus eingeschaltet werden. Ein Vorteil gegenüber den schrittweitengesteuerten Einschrittalgorithmen ist, daß die Effektivität der entsprechenden Zweischrittalgorithmen von der Wahl des Koordinatensystems nicht abhängt.

Da weiterhin der Adaptionsschritt aus zwei relativ unabhängigen Komponenten, dem Primärterm $\gamma\mathbf{s}$ und dem Trendterm $r[n] \cdot (\boldsymbol{w}[n-1] - \boldsymbol{w}[n-2])$, additiv zusammengesetzt ist, ist die Gefahr eines „Festfressens" schon vom Ansatz her weitgehend ausgeschaltet.

Bei der Diskussion der Stabilitätsbedingung ist eine Besonderheit zu beachten. Im Primärterm ist explizit (bei den Gradientenverfahren) oder implizit (bei den Suchverfahren) ein Anteil $-\gamma_0 \boldsymbol{w}[n-1]$ enthalten. Die charakteristische Gleichung hat somit die Gestalt

$$\mu^2 - (1 - \gamma_0 + r) \cdot \mu + r = 0 . \tag{3.4}$$

Durch elementare Rechnung findet man leicht die beiden Bedingungen

$$r < 1 \quad \text{und} \quad r - \gamma_0 > -1 . \tag{3.5}$$

Danach kann sich r asymptotisch beliebig an $+1$ annähern, an -1 hingegen nicht. Hieraus folgt für die α_i:

$$\alpha_1, \alpha_2 \geqq 0 , \tag{3.6a}$$

$$\alpha_1 \geqq \alpha_2 + \alpha_3 , \tag{3.6b}$$

$$\alpha_3 = \alpha_3(\gamma_0) > 0 . \tag{3.6c}$$

Um den Algorithmus zweiter Stufe so wirksam als möglich zu machen, wird in (3.6b) meist das Gleichheitszeichen gesetzt.

Das weiteren sollte gelten

$$\alpha_2 > \alpha_3 . \tag{3.6d}$$

Damit sind negative $r[n]$ grundsätzlich möglich. Sie sind in vielen Fällen sogar erwünscht, z.B. dann, wenn es gilt, eine Adaptionsrichtung, die sich aktuell als ungünstig erweist, rasch abzubremsen. Bei Erreichen eines relativen Optimums in der aktuellen Abstiegsrichtung schießt der Algorithmus mehr oder weniger über das Ziel hinaus. Mit negativem $r[n]$ liefert der homogene Term einen eigenen Beitrag, $\boldsymbol{w}[n]$ in die Nähe des Optimums zurückzubringen.

Das Glättungsverhalten des Algorithmus, das vor allem bei Adaptionsproblemen besonderes Interesse verdient, wird wesentlich von α_1, das die Funktion einer Zeitkonstanten hat, bestimmt.

Eine gewisse Einsicht in die Dynamik der Zweischrittadaption ist durch das Studium des homogenen Prozesses zu gewinnen:

$$\boldsymbol{w}[n] = \boldsymbol{w}[n-1] + r[n]\,(\boldsymbol{w}[n-1] - \boldsymbol{w}[n-2])\,, \tag{3.7}$$

$$r[n] = (1-\alpha_1)\,r[n-1] + \alpha_2 \operatorname{sign} \Delta Q[n] + \alpha_3\,. \tag{3.8}$$

Unter den Voraussetzungen

$$r[0] = 0\,,$$

$$\alpha_1 = \alpha_2 + \alpha_3 = \alpha\,, \tag{3.9}$$

und

$$\Delta Q[k] > 0 \quad \text{für } k = 1, \dots n$$

ergibt sich

$$\boldsymbol{w}[n] = \boldsymbol{w}[0] + \left\{\sum_{\nu=1}^{n} \prod_{\mu=2}^{\nu} \left(1 - (1-\alpha)^{\mu-1}\right)\right\} (\boldsymbol{w}[1] - \boldsymbol{w}[0])\,. \tag{3.10}$$

Da die Reihe

$$S_\nu = \sum_{\mu=2}^{\nu} (1-\alpha)^{\mu-1} \tag{3.11}$$

konvergiert, bleibt auch das Produkt in (3.10) für $\nu \to \infty$ endlich. Der zweite Term in (3.10) vermag somit unter der Annahme $\boldsymbol{w}[1] - \boldsymbol{w}[0] \neq 0$ jeden vorgegebenen Wert anzunehmen. Mithin erreicht der homogene Prozeß stets ein relatives Optimum. Da seine Richtung durch $\boldsymbol{w}[1] - \boldsymbol{w}[0]$ festgelegt ist, ist es i. allg. nicht möglich, zum globalen Optimum zu gelangen.

Eine Besonderheit des betrachteten zweistufigen Zweischrittalgorithmus besteht mithin darin, daß er die Stabilitätsbedingung (3.5) erfüllt und doch eine beliebig lange Strecke im Zustandsraum der $\boldsymbol{w}[n]$ zurückzulegen vermag.

4. Einsatz von Zweischrittalgorithmen bei schlecht konditionierten Problemen

Schlecht konditionierte Probleme im Sinne der numerischen Mathematik sind dadurch gekennzeichnet, daß die Empfindlichkeit gegenüber Parameteränderungen nach Ort und Richtung extreme Unterschiede aufweist. Zu den schlecht konditionierten Problemen gehören auch solche mit schluchtförmigem Kostenfunktional. In solchen Fällen eröffnen die Zweischrittalgorithmen eine einfache Möglichkeit zur Konvergenzbeschleunigung. Die Voraussetzungen an die Art des Kostenfunktionals sind dabei sehr schwach. Sie betreffen im wesentlichen Unimodalität und eine gewisse Glattheit. Letztere Forderung wird weiter unten genauer erläutert.

Betrachten wir ein Gradientenverfahren. Die maximale Schrittweite wird von der Steilheit und der jeweiligen Entfernung zum Optimum bestimmt. Allgemeine Aussagen gibt es nur für Probleme mit höchstens quadratischer Ordnung, d. h. wenn $|J(\boldsymbol{w})|$ nicht schneller als $|\boldsymbol{w} - \boldsymbol{w}^*|^2$ wächst. Aufgrund der Schrittweitenbeschränkung kommt der Einschrittalgorithmus, einmal auf der Sohle der Schlucht angelangt, nur noch langsam voran. Dies wird als starke Unterrelaxation bezeichnet [2]. Im Zweischrittalgorithmus bleibt $\Delta Q[n]$ über mehrere Takte positiv, $r[n]$ nimmt asymptotisch den Wert Eins an und der Abstieg wird beschleunigt.

Es lassen sich Voraussetzungen lokaler Natur dafür angeben, daß der Zweischrittalgorithmus effektiv arbeitet. Es wird erstens angenommen, daß das adaptive System noch relativ weit vom Optimum entfernt ist, so daß die Ungleichung

$$|\boldsymbol{w}[n-1] - \boldsymbol{w}[n-2]| \ll |\boldsymbol{w}[n-1] - \boldsymbol{w}^*| \tag{4.1}$$

erfüllt ist. Die Trajektorie verlaufe hinreichend glatt, so daß die Annahme

$$(\boldsymbol{w}[n-1] - \boldsymbol{w}^*)^T (\boldsymbol{w}[n-2] - \boldsymbol{w}^*) > |\boldsymbol{w}[n-1] - \boldsymbol{w}^*|^2 \tag{4.2}$$

gemacht werden kann. Dies ist gleichzeitig eine Bedingung an die Glattheit des Kostenfunktionals. Bei Stabilität des homogenen Teils im Algorithmus gemäß (3.1) und Unimodalität von $J(\boldsymbol{w})$ arbeitet der Zweischrittalgorithmus stabil. In großen Entfernungen zum Optimum, bezogen auf die Schrittweite, gilt die Abschätzung

$$|\boldsymbol{w}[n] - \boldsymbol{w}^*|_{\text{Einschritt}} > |\boldsymbol{w}[n] - \boldsymbol{w}^*|_{\text{Zweischrift}}\,, \tag{4.3}$$

wobei $\boldsymbol{w}[n-1]$ für beide Fälle als gleich angenommen wird.

Die Konvergenz läßt sich demnach bei starker Unterrelaxation beschleunigen. Der Trendterm stellt sich weitgehend unabhängig von der speziellen Wahl des Koordinatensystems ein. Beim Einschrittverfahren ist diese Invarianz prinzipiell auch erreichbar. Allerdings bedeutet dies einen relativ hohen Aufwand, da in der zweiten Stufe eine Matrix adaptiv zu bestimmen ist [4] (vgl. (1.11)).

Ein kritischer Punkt des Zweischrittverfahrens ist seine Anfälligkeit gegen Überschwingen beim Wechsel vom Übergangsbetrieb ($\Delta Q > 0$) zum ruhigen Betrieb ($\langle \Delta Q \rangle \approx 0$). Durch geeignete Wahl der α_i wird dieses Überschwingen hinlänglich unterdrückt. Entsprechende quantitative Aussagen, die jeweils für bestimmte Problemklassen gelten, ergeben sich durch Auswertung der Simulationsergebnisse.

5. Simulationsergebnisse

Die Simulationen dienten zwei Zielen. Erstens sollte die Frage, wie die Parameter α_i in Abhängigkeit von der Problemklasse zu wählen sind, geklärt werden. Zum zweiten sollte ein Leistungsvergleich zwischen Ein- und Zweischrittalgorithmen erfolgen.

Es werden ausschließlich schlecht konditionierte Probleme betrachtet, weil sonst keine echte Notwendigkeit besteht, die herkömmlichen Verfahren zu modifizieren.

Abb. 3 stellt die Einschrittgradientenmethode

$$\boldsymbol{w}[n] = \boldsymbol{w}[n-1] - \gamma \nabla_{\boldsymbol{w}} J(\boldsymbol{w}) \tag{5.1}$$

der Zweischrittgradientenmethode

$$\boldsymbol{w}[n] = \boldsymbol{w}[n-1] + r[n]\,(\boldsymbol{w}[n-1] - \boldsymbol{w}[n-2]) - \gamma \nabla_{\boldsymbol{w}} J(\boldsymbol{w}) \tag{5.2}$$

$$r[n] = (1-\alpha_1)\, r[n-1] + \alpha_2 \operatorname{sign}(\Delta Q(n-1)) + \alpha_3 \tag{5.3}$$

anhand des zweidimensionalen quadratischen Problems

$$J(\boldsymbol{w}) = (\boldsymbol{w} - \boldsymbol{w}^*)^T A (\boldsymbol{w} - \boldsymbol{w}^*) \to \min$$

gegenüber. A ist eine (2×2)-Matrix mit stark voneinander verschiedenen Eigenwerten. Die Anzahl der Schritte ist für alle Trajektorien gleich. Aufgrund der schlechten Kondition des Problems muß ein relativ kleiner Koeffizient γ gewählt werden. Am Boden der Schlucht, der nach wenigen Schritten erreicht ist, geht der Abstieg extrem langsam weiter. Es ist zu erkennen, daß das Zweischrittverfahren den Abstieg erheblich beschleunigt.

Das Verhalten bei einem Problem mit einer Beschränkung — die untere Halbebene ist nicht zugänglich — wird in Abb. 4 und 5 verdeutlicht. Bei der vorgegebenen Anzahl von 40 Schritten erreicht das Zweischrittverfahren (Abb. 5) das Optimum in jedem Falle, das Einschrittverfahren (Abb. 4) jedoch nur in einer der Simulationen.

Als Empfehlung für die α_i ergibt sich für diese Problemklasse:

$$0{,}3 \leqq \alpha_1 \leqq 0{,}6\,, \tag{5.4}$$

$$1 \leqq \alpha_2/\alpha_3 \leqq 4\,. \tag{5.5}$$

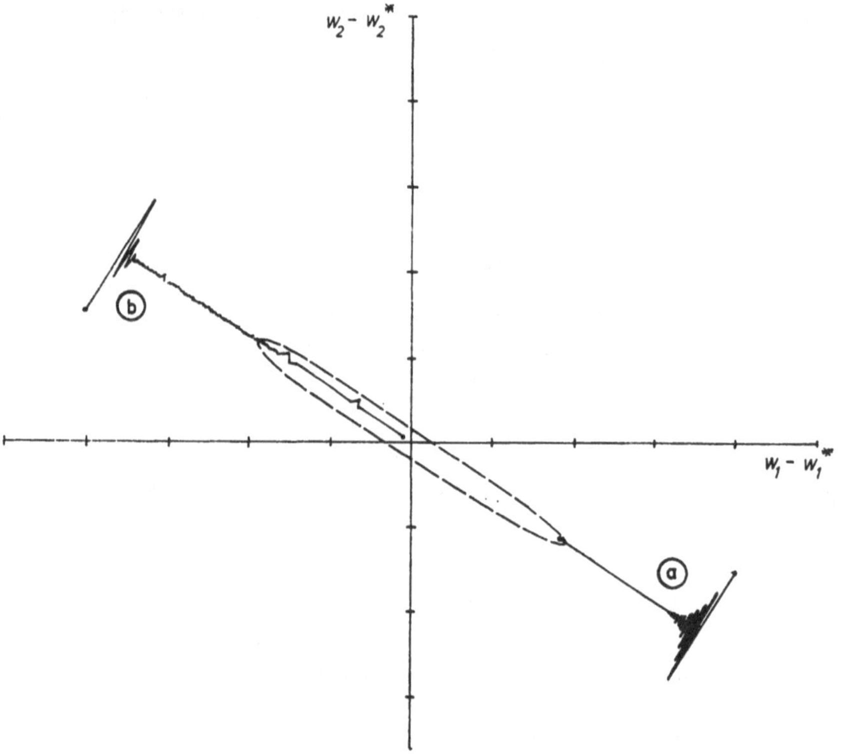

Abb. 3. Trajektorien der iterativen Optimierung (Einschrittgradientenverfahren (a) und Zweischrittgradientenverfahren (b)). Die eingezeichnete Ellipse stellt hier wie auch in Abb. 4 und 5 eine Kurve $J(\boldsymbol{w}) = \text{const.}$ dar.

Es sollte erwähnt werden, daß diese Aussagen nicht an die Klasse der quadratischen Kriterien gebunden sind. Dies wird durch ein weiteres Simulationsexperiment mit dem Kriterium

$$J(\boldsymbol{w}) = w_1^2 + w_2^2 + w_3^2 + 10[(w_4 - w_3^2)]^2 \tag{5.6}$$

verdeutlicht. Auf der Grundlage der Gradientenmethode werden drei Algorithmen untersucht, das Einschrittverfahren (nach (5.1)), das schrittweitengesteuerte (zweistufige) Einschrittverfahren

$$\boldsymbol{w}[n] = \boldsymbol{w}[n-1] - \bar{\gamma}[n] \cdot \nabla_w J(\boldsymbol{w}[n-1]), \tag{5.7}$$

$$\bar{\gamma}[n] = (1 - \alpha_1)\,\bar{\gamma}[n-1] + \gamma \cdot \alpha_2 \operatorname{sign}(\Delta Q[n-1]) + \gamma \cdot \alpha_3 \tag{5.8}$$

und das zweistufige Zweischrittverfahren ((5.2) und (5.3)).

Die Ergebnisse der Simulation sind in Abb. 6, 7 und 8 dargestellt. Die Empfehlungen für die Wahl der α_i ließen sich bestätigen. In der Konvergenzgeschwindigkeit sind die beiden zweistufigen Verfahren dem einfachen Einschrittverfahren überlegen. Das Zweischrittverfahren erzeugt eine relativ glatte Trajektorie im Vergleich zum zweistufigen Einschrittverfahren. Damit deutet sich seine bessere Stabilität an.

Die Optimierungsbeispiele dienen der Illustration. Das eigentliche Anwendungsgebiet der Zweischrittalgorithmen liegt jedoch innerhalb der Adaption. Es wurde eine Polynomapproximation untersucht.

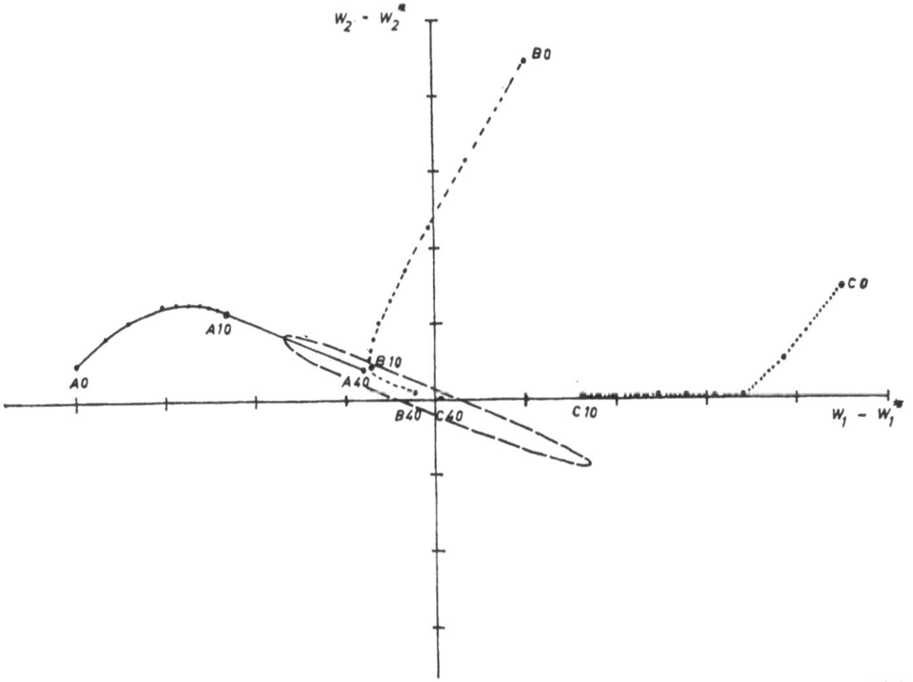

Abb. 4. Trajektorien nach dem Einschrittverfahren in 40 Takten (A0, B0, C0 — Beginn; A40, B40, C40 — Ende)

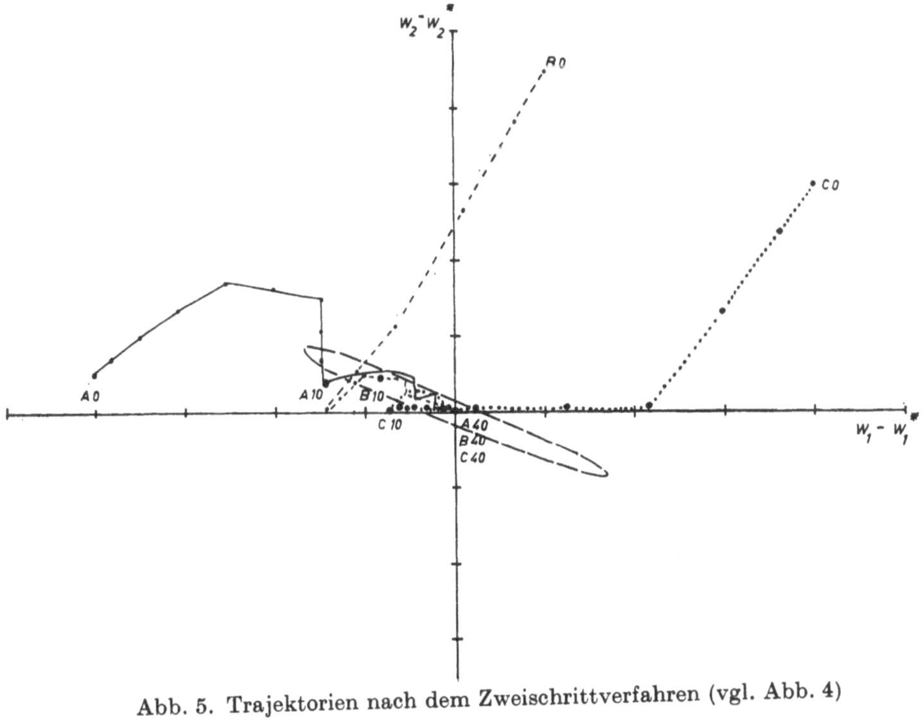

Abb. 5. Trajektorien nach dem Zweischrittverfahren (vgl. Abb. 4)

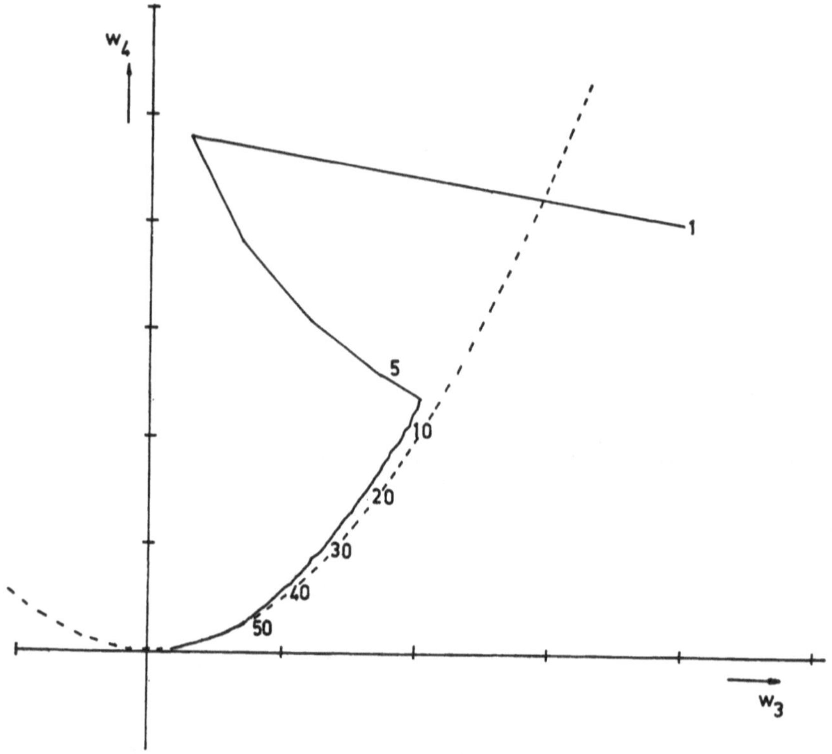

Abb. 6.

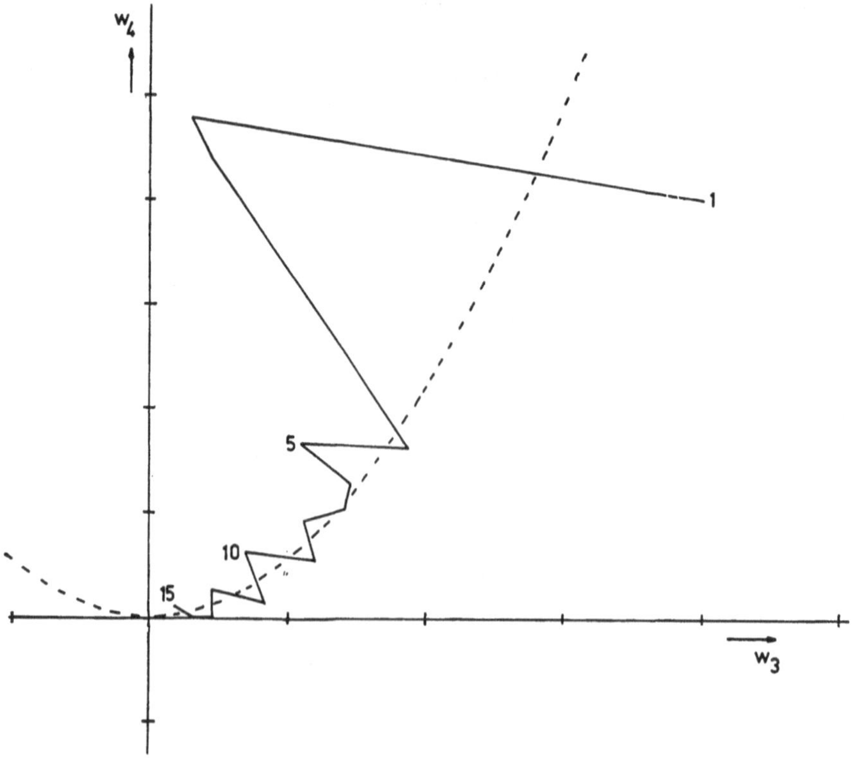

Abb. 7.

Abb. 6. Verhalten des Einschrittgradientenverfahrens bei schluchtförmiger Kostenfunktion

◀ Abb. 7. Verhalten des schrittweitengesteuerten Einschrittgradientenverfahrens (die Parabel stellt den Verlauf der Schluchtsohle dar)

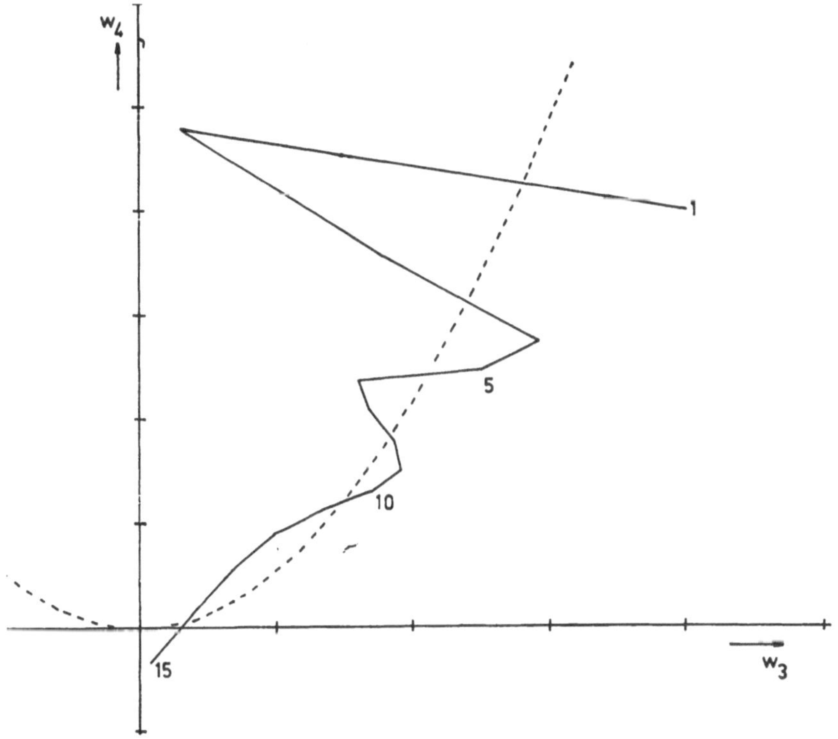

Abb. 8. Verhalten des zweistufigen Zweischrittgradientenverfahrens

Wie man sich leicht überlegen kann, sind solche Regressionsprobleme in der Mehrzahl der Fälle schlecht konditioniert, so auch die vorliegende Aufgabe. Die in Abb. 9 durch Meßpunkte dargestellte funktionale Abhängigkeit[1]) soll durch ein Polynom dritter Ordnung mit minimaler quadratischer Abweichung wiedergegeben werden.

Es werden drei Algorithmen hinsichtlich der Adaptionsgeschwindigkeit verglichen, der einstufige Einschrittalgorithmus (AGEINF), der zweistufige Zweischrittalgorithmus (AGZWEI) und der optimale Algorithmus nach Albert und Gardner (AGALGA). Jeder der Algorithmen wurde je sechsmal mit geänderter zufälliger Auswahl der Meßpunkte simuliert. Mit einer Periode von jeweils 31 Schritten wird der gleitend berechnete mittlere Abstand Q_m ausgegeben. In Abb. 10 sind jeweils Minimum und Maximum von Q_m für jede Periode dargestellt. Die Differenzen zwischen Minimum und Maximum sind genügend klein, so daß Aussagen von ausreichender Signifikanz möglich sind. Der Zeitmaßstab wurde für AGZWEI und AGALGA entsprechend der im Vergleich zu AGEINF erhöhten Rechenzeit gedehnt. Bei dieser kombinierten Be-

---

[1]) Die Meßdaten stellte freundlicherweise das Institut für Physik der Werkstoffbearbeitung der AdW zur Verfügung.

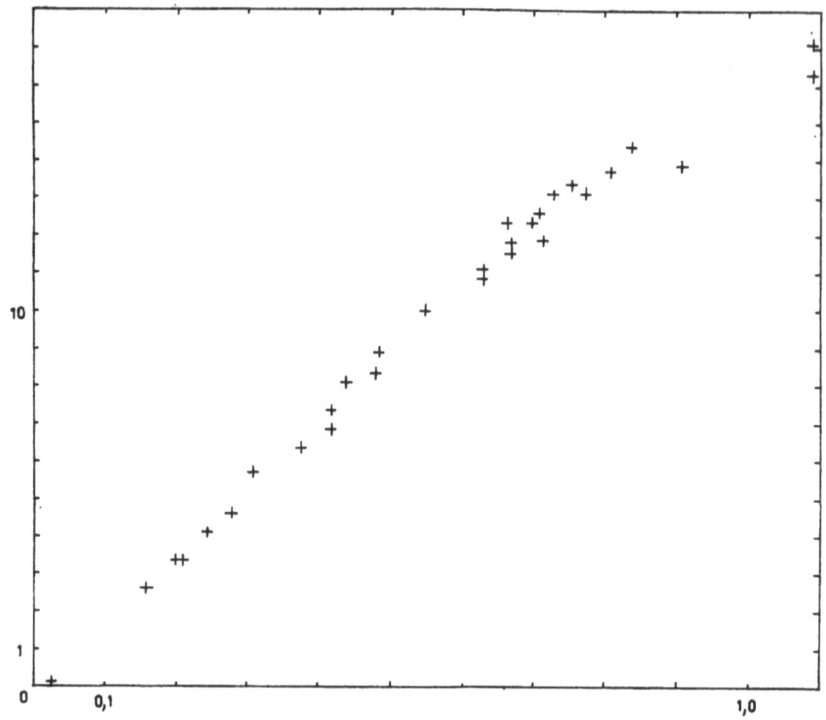

Abb. 9. Meßpunkte einer Eichkurve (gemessene Temperaturabhängigkeit einer Thermospannung)

wertung von Genauigkeit und Rechenzeit sind die Vorteile des Zweischrittverfahrens besonders nach 4 bis 8 Perioden deutlich erkennbar. Der Vergleich zum optimalen Algorithmus würde bei höherem Grad N des Regressionspolynoms noch günstiger ausfallen, da der Aufwand zur Ermittlung der Matrix Γ quadratisch mit N zunimmt.

Bei Adaptionsproblemen sollte die Zeitkonstante α_1 klein genug sein, um eine hinreichende Glättung des Einflusses der verschiedenen Meßwerte innerhalb der Stichprobe zu gewährleisten. Im vorliegenden Falle ergab die Abschätzung $\alpha_1 \approx 0{,}03$, was durch die Simulation bestätigt werden konnte.

Die günstigen Eigenschaften von Zweischrittalgorithmen lassen sich auch zur beschleunigten Belehrung von Klassifikatoren verwenden.

6. Schlußbetrachtung

Als Folge der Zweistufigkeit adaptieren die betrachteten Algorithmen in einem doppelten Sinne. Die Zweischrittigkeit legt die spezielle Eigenschaft der Adaption der zweiten Stufe als selbsttätige Trendbewertung fest. Daraus folgt der nichtstationäre Charakter der Algorithmen, der im wesentlichen von der Fähigkeit herrührt, zwischen verschiedenen Betriebszuständen während der Adaption der ersten Stufe umschalten zu können. Deshalb ist es auch nicht möglich, die Zweischrittalgorithmen aus einem Gütekriterium herkömmlicher Gestalt direkt abzuleiten.

Diese sind als Scharmittelwerte definiert und somit nicht fein genug, das lokale Verhalten der Algorithmen differenziert zu beeinflussen.

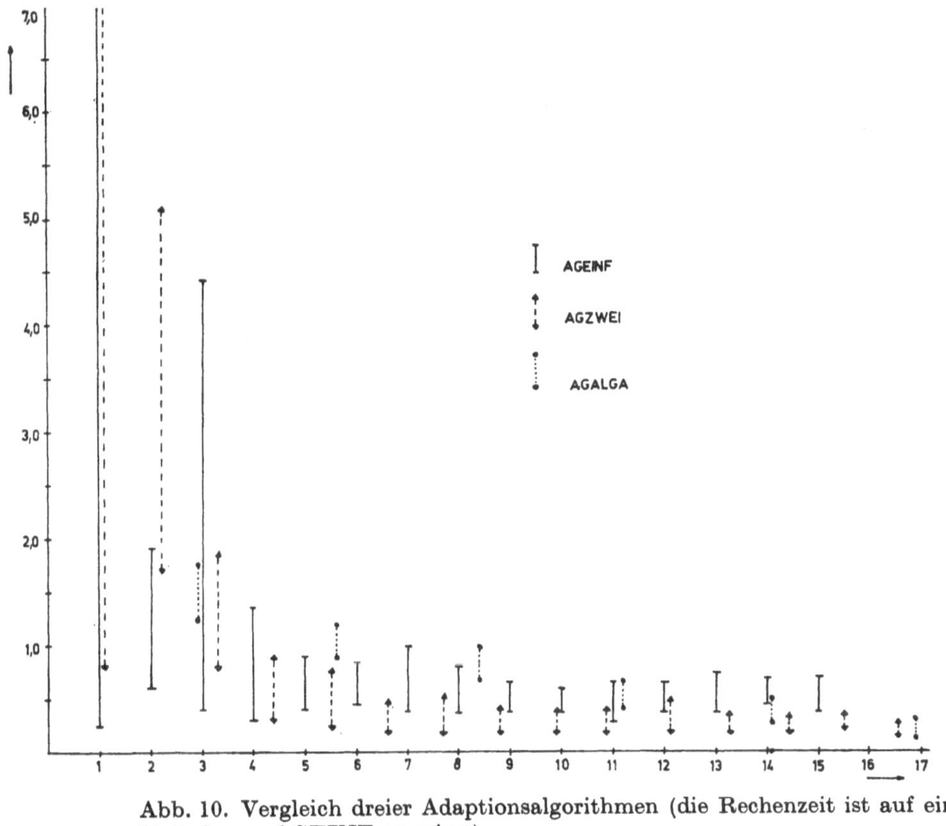

Abb. 10. Vergleich dreier Adaptionsalgorithmen (die Rechenzeit ist auf eine Periode von AGEINF normiert)

Die Betrachtungen zum Verhalten des homogenen Teils der Zweischrittalgorithmen zeigten, daß sie global stabil und gleichzeitig in der Lage sind, eine beliebig große Entfernung zwischen Anfangs- und Endzustand zurücklegen zu können.

Trotz einer Reihe präziser Aussagen zum Verhalten einzelner Teile der Algorithmen fehlt eine umfassende Theorie. Deshalb werden Rechnersimulationen zur Demonstration der Wirkungsweise der Algorithmen ausgeführt, eine Vorgehensweise, die z.B. in der Monographie von Solodownikow [14] empfohlen wird.

Die mehrstufigen Mehrschrittalgorithmen bieten zahlreiche Möglichkeiten für weitere Untersuchungen. So wären u. a. die Fragen zu prüfen, inwieweit die Adaption der α_i in einer dritten Stufe sinnvoll sein könnte, ob der Einsatz von Zufallssuchprozessen Vorteile bietet und welche Klassen nichtstationärer Optimierungsprobleme erfolgversprechend behandelt werden können.

Beim praktischen Einsatz der Mehrschrittalgorithmen spielt die Frage der geeigneten Geräterealisierung eine wichtige Rolle. Die von Köhler und Franke [15] entwickelten adaptiven Strukturen bieten günstige Voraussetzungen in dieser Hinsicht.

Literatur

[1] *Zypkin, J. Z.*, Adaption und Lernen in kybernetischen Systemen. VEB Verlag Technik, Berlin 1970.

[2] *Göpfert, A.*, Mathematische Optimierung in allgemeinen Vektorräumen. B. G. Teubner Verlagsgesellschaft, Leipzig 1973.

[3] *Kantorowitsch, I. W., G. P. Akilow,* Funktionalanalysis in normierten Räumen. Akademie-Verlag, Berlin 1964.
[4] *Albert, A. E., L. A. Gardner,* Stochastic Approximation and Nonlinear Regression. Research Monograph No. 12, MIT Press, Cambridge (Mass.) 1967.
[5] *Цыпкин Я. З.*, Квазиоптимальные алгоритмы обучения. Автоматика и телемеханика 6/1973, 31—40.
[6] *Kesten, H.*, Accelerated stochastic approximations. Ann. Math. Statist. **29** (1958), 41—59.
[7] *Невельсон М. Б., Р. З. Хасьминский*, Адаптивная процедура Роббинса-Монро. Автоматика и телемеханика 10/1973, 71—83.
[8] *Цыпкин Я. З.*, Основы теории обучающихся систем. Наука, Москва 1970.
[9] *Höge, H.*, Stochastic algorithms in system identification. Int. J. Control **17** (1973), 1121—1128.
[10] *Saridis, G. N.*, Learning Applied to Successive Approximation Algorithms. IEEE Trans. **SSC-6** (1970), 97—103.
[11] *Растригин Л. А.*, Системы экстремального управления. Наука, Москва 1974.
[12] *Stoer, J., R. Bulirsch*, Einführung in die Numerische Mathematik II. Springer-Verlag, Berlin 1973.
[13] *Юдин Д. Б.*, Методы количественного анализа сложных систем. Изв. АН СССР, техн. киберн. 1/1965, 3—13.
[14] *Solodownikow, W. W.*, Instationäre und nichtlineare Systeme. VEB Verlag Technik, Berlin 1974.
[15] *Köhler, L., A. Franke*, Informationsverarbeitender Speicher als adaptiver elektronischer Meßgrößenwandler. Diss., TH Karl-Marx-Stadt, 1973.

Kurzfassung

Die Leistungsfähigkeit von Iterationsalgorithmen läßt sich durch verschiedene Erweiterungen verbessern. In Mehrstufenalgorithmen wird eine Hierarchie von Verstellregeln gebildet. Hierzu gehören insbesondere die schrittweitengesteuerten Algorithmen. In Mehrschrittalgorithmen werden zur Berechnung des Zustandes $\boldsymbol{w}[n]$ eines adaptiven Systems neben dem Zustand $\boldsymbol{w}[n-1]$ auch weiter zurückliegende Zustände $\boldsymbol{w}[n-i]$ explizit berücksichtigt. Der vorliegende Beitrag befaßt sich mit zweistufigen Zweischrittalgorithmen und untersucht deren Eignung zur Lösung schlecht konditionierter Adaptions- und Optimierungsprobleme.

Abstract

The efficiency of iteration algorithms can be increased by several methods. In multilevel algorithms a hierarchy of adjusting rules is built up. Particularly, all algorithms with a special adjustment of the step size belong to this type. In multistep algorithms the state $\boldsymbol{w}[n]$ of an adaptive system is calculated using explicitly not only the state $\boldsymbol{w}[n-1]$, but also some states $\boldsymbol{w}[n-i]$ with $i > 1$. In this paper, two-level two-step algorithms are considered. It is shown that they can be used for the accelerated solution of ill-conditioned adaptation and optimization algorithms.

Резюме

Как известно, можно улучшить работоспособность итеративных алгоритмов при помощи подходящих расширений. В многостепенных алгоритмах построена хиерархия настраивающих правил. Примерами являются алгоритмы с управлением длиной шагов. В многошаговых алгоритмах при определении состояния $\boldsymbol{w}[n]$ адаптивной системы учитиваются в явной форме не только $\boldsymbol{w}[n-1]$, а тоже некоторые

$w[n-i]$ $(i > 1)$. Настоящая статья занимается двухстепенными, двухшаговыми алгоритмами и исследует их применимость при ускоренном решении овражных проблем адаптации и оптимизации.

(Eingang: erste Fassung am 17. 7. 1975,
vorliegende Fassung am 13. 5. 1976)

Anschrift des Verfassers:

Dr. K. Fritzsch
Zentralinstitut für Kybernetik und
Informationsprozesse der AdW
1199 Berlin
Rudower Chaussee 5
DDR

Elektronische Informationsverarbeitung und Kybernetik
EIK 13 (1977) 1/2, 79 – 196

A Formalization of the Transformation of Nested Declarations[1])

By *Mauro Bert*[2])

1. Introduction

Written in the so-called Backus normal form, the formulae, that define ALGOL 60, only give a partial characterization of ALGOL syntax, since there are some synctactical rules which remain unformalized in the system and which rules have been given in the explanations and comments accompanying the formulae; see Naur [9], Caracciolo di Forino [3], Floyd [5]. These are, for instance, the rules that all the identifiers used in a block are to be declared, that no identifier can be declared twice in one block, and that the declaration of an identifier given within the 'begin' block is valid only inside that block.

The Backus normal form is equivalent to a context-free grammar (CFG), and it is for this reason that ALGOL 60 has been considered a context-free language. However, although it is true that the set of formulae which appear in [9] do in fact define a context-free language (CFL), this language is not really ALGOL 60, but differs from it by a set of strongly restricting conditions. The nature of these restricting conditions is tightly connected with the peculiar character of the declarations.

Let us now take an example of an ALGOL block, whose semantics will be studied by pointing out the problem of declarations:

begin real a, **real** b, **integer** c

$a := b + c;$ (i)

begin real c;

$b := a \times c;$ (ii)

end

$a := 0;$ (iii)

begin integer b;

$b := c;$ (iv)

end

end

The declarations serve to define certain properties of the identifiers of the program. A declaration for an identifier is valid for one block. Outside of this block the particular identifier may be used for other purposes. In our example within statement (i) a and

[1]) These results were obtained in the author's doctoral dissertation, supported by the Italian C.N.R. (Fellowship n. 209.1.15).

[2]) Istituto di Scienze dell'Informazione, Università di Torino (Director: Prof. Ing. V. Mauro).

b are real, and c integer; within statement (ii) b, a, c are real; within statement (iii) a is real; within statement (iv) b and c are integer.

It is well known that the semantics of a programming language A can be described in the following way: for every string α of A we exhibit a string β of a second language B, whose semantics is supposed to be known. Since A usually contains an infinite number of strings, for this association to be effective, it is necessary to have a "finite device" which for every $\alpha \in A$ produces the string $\beta \in B$ equivalent to α. But then the description of the semantics of A is nothing further than the definition of a mapping of A into B or, as it is usually called, the compiler which translates A into B.

Therefore defining the semantics of declaration is equivalent to the problem of finding a translation scheme which is capable to transform an ALGOL-like language into a language without a nested structure of declarations.

2. Basic Definitions

A particular family of mappings between languages based upon their grammars, T_1-semantics, was defined by Petrone [10, 11]. T_1-semantics are a generalization of the "syntax directed translations" by Lewis and Stearns [8].

Before defining T_1-semantics, first we give some basic definitions. We will write T_Σ for the set of strees, whose nodes are labeled with elements of the alphabet Σ (Σ trees).

Definition 2.1. For $t \in T_\Sigma$, the *production tree* of t, denoted $\mathcal{P}(t)$, is the unique $(\Sigma \cup \Sigma^2 \cup \Sigma^3 \cup \cdots \cup \Sigma^n \cup \Sigma^{n+1})$ tree which is structurally equivalent to t (thus having the same frontier as t) and which for every internal node has the label $(\sigma, \sigma_1, \ldots, \sigma_n)$ if the corresponding node of t has label σ and successors labeled $\sigma_1, \ldots, \sigma_n$. We will write $T_\mathcal{P}(G)$ for the set of production trees of the derivation trees of a context-free grammar $G = (N, Z, P, S)$.

Definition 2.2. Let $\mathcal{B} = \{Y, O_1, \ldots, O_k\}$ be an algebra where Y (the carrier) is an alphabet of output symbols and the O_i are operators on n_i operands. A pair of maps, $A\colon P \to \{O_1, \ldots, O_k\}$ (where P is the set of productions of G) and $A_0\colon Z \to Y$ determine a homomorphism μ from $T_\mathcal{P}(G)$ into $\mathcal{B}$ by:

a) If t is a single node labeled σ, then $\mu(t) = A_0(\sigma)$.

b) If t has the form of Figure 1, then $\mu(t) = A(\sigma, \sigma_1, \ldots, \sigma_n)\big(\mu(t_1), \mu(t_2), \ldots, \mu(t_n)\big)$.

The homomorphism μ, above defined, is called *syntax-directed mapping* of L, where L is the language generated by G.

Each class M of mappings will be obtained in correspondence to a certain class T of operators, which is subject to one condition: it is closed under composition, i.e., the result of applying one operator to another is still an operator of the class T.

Definition 2.3. The map A of the above definition is called a *T-semantics* on base G if an operator O_i on n operands, of class T, is given for every production rule $B \to \alpha$, where α is a word of lenght n.

Let us now introduce the *class* T_1. T_1 is the closure by composition of the binary concatenation operator $C(x_1, x_2)$ and of the constant operators $K_w(x_1)$. By definition, the operator $C(x_1, x_2)$ when applied to a pair of strings w_1 and $w_2 \in Y^*$, gives string w_1w_2; the operator $K_w(x_1)$ is, for any string x_1 of Z^*, the fixed string w of Y^*.

Examples of unary operators of class T_1 are the unary operator $K_e(x)$, e is the empty string, the identity operator $I(x_1)$ defined by $I(x_1) \equiv C(x_1, K_e(x_1))$ and the operator $C\big(x_1, C(C(K_w(x_1), x_1), x_1)\big)$ that when applied, for instance to string *baa*, yields string *baawbaabaa*.

Because of the associative property of concatenation, for each unary operator of T_1, there exists a concise representation that we shall call the *representing word* (rw).

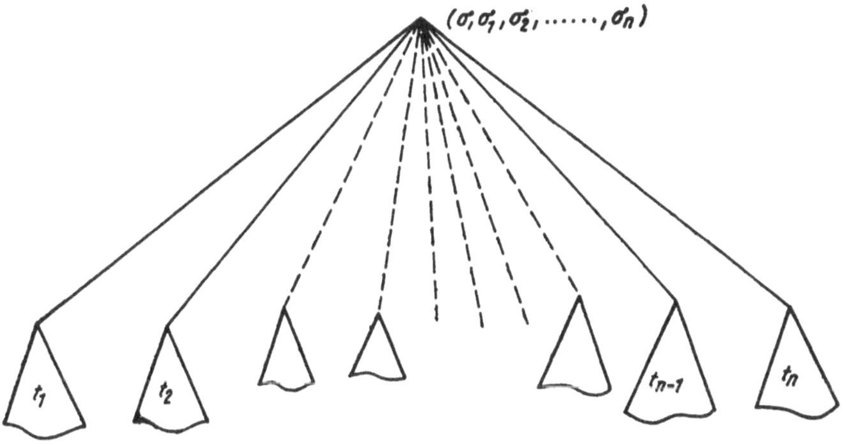

Fig. 1

The representing word is obtained by applying the operator to a conventional fixed one letter string x_1 not belonging to Y^* and denoting concatenation by juxtaposition. The representing words for the above mentioned operators are e, x_1, $x_1wx_1x_1$, respectively. The same concise representation can be given for the n-ary operators of class T_1. The operator is applied to arguments $x_1, \ldots, x_n$ considered as strings in the alphabet $X = \{x_1, \ldots, x_n\}$. The representing word of the more general n-ary operator of class T_1 is a word in the alphabet $X \cup Y$.

Let us now consider a simple example of mapping of class T_1 for the grammar defined by the following production rules:

1) $S \rightarrow a$
2) $S \rightarrow bRS$
3) $R \rightarrow cS$

We shall define the operators O_i through their representing words, and suppose $\mu(t) = t$ for any terminal t.

Example. Let μ_1 be defined by the T_1-semantics S_1:

$$S_1 = \{O_1(x_1) \equiv m; \quad O_2(x_1, x_2, x_3) \equiv x_1x_2cx_3d; \quad O_3(x_1, x_2) \equiv nx_2p\} .$$

If we denote correspondence under μ_1 by the arrow "$\rightarrow$", we obtain that

$$a \rightarrow m$$

$$bcbcaaa \rightarrow O_2(b, O_3(c, O_2(b, O_3(c, O_1(a)) , O_1(a))), O_1(a)) =$$
$$= O_2(b, O_3(c, O_2(b, nmp, m)), m) = O_2(b, O_3(c, bnmpcmd), m) =$$
$$= bnbnmpcmdpcmd$$

3. Semantics of Class T_2 and Nested Declarations

We will now consider another class of operators: T_2.

Definition 3.1. *Class* T_2 is obtained from T_1 by introducing a new ternary operator $\lambda(x_1, x_2, x_3)$ defined as follows: the argument x_3 is first evaluated, then, each occurrence of x_2 in the string which is the value of x_3, is substituted by x_1.

In our definition the arguments x_1 and x_2 are not expressions, therefore our definition of class T_2 is more restricted than the analogous definition in [11].

There is an evident difference in the power between T_1 and T_2-semantics. The first one is a translation scheme which can do only text transformations (change the position of terminals, code, duplicate and erase terminals). The second one, being able to manage the problem of indentifiers declaration and their use, is more powerful; for this reason T_2-semantics represent a class of translation schemes more complete and closer to real situations.

We will see, in the following, a natural extension of T_2-semantics obtained from GSDT's augmented by the λ-operator.

We will tackle the problem of nested declarations by making use of the substitution operator λ. To make an example, if we apply λ to the following three arguments:

$$x_1 = \text{type} \qquad x_2 = \text{identifier} \qquad x_3 = \text{block}$$

each occurrence of the identifier x_2 in the block will be substituted by the type associated to same in the declaration.

Note, furthermore, that our λ is a functional operator, i.e. it can be applied when another λ or an operator of class T_1 has already been applied to argument x_3. This property allows us to respect the scope of the variables which are declared.

Example. Suppose we have the following program written in an ALGOL-like language:

```
begin real b;
  b = 0;
  begin boolean b;
    b = false;
  end
end
```

This program contains two nested blocks. We will show that the meaning of the previous program may be expressed by the following functional expression, made by a set of operators associated to the production rules of the grammar:

$$\lambda(\mathbf{real}, b, O_1(b = 0, \lambda(\mathbf{boolean}, b, b = \text{false})))$$

where O_1 is a binary operator of class T_1. Because of the structure of the functional expression, the boolean type will be assigned to "b" of the second statement and the real type to "b" of the first statement.

It is clear now what importance the λ-operator has in the compiling of block languages: it allows us to eliminate the nesting, respecting the rules given by the scope definition.

It is important to note that problems of declarations nesting also exist in other languages besides ALGOL 60, such as PL/I and ALGOL 68.

We have given the name "λ" to the substitution operator, purposely. In fact, our substitution operator has a close ***relationship to the λ-calculus***. (The reader can find the fundamental notions about λ-calculus in [4]). In particular, our operator is connected to the β-rule of the λ-calculus, which is an elementary substitution rule. Let M, L be two formulae of the λ-calculus. The writing $L =_{\mathrm{df}} \mathcal{L}[x]$ points out the possible dependence of L on every free occurrence of variable x. Furthermore we denote with $\mathcal{L}[M]$ the simultaneous substitution of every free occurrence of x in $\mathcal{L}[x]$ by M. The β-rule works in the following way:

$$\lambda x \mathcal{L}[x]\, M \xrightarrow{\beta} \mathcal{L}[M]\,.$$

The equivalent expression with our λ-operator is

$$\lambda(M, x, \mathcal{L}[x]) = \mathcal{L}[M] .$$

An important comment, however, needs to be made: in the application of the β-rule to the problems of declarations of identifiers we need not suppose that M is, in turn, *a* λ-formula; that is, the expression M, that defines the type of the identifier x, need not be, in turn, a generalized program, involving in turn possible substitution operations. The λ-operator is then, certainly, a particular case of the more general substitution rule of the λ-calculus.

We can, therefore, consider the ***subclass of the λ-calculus*** obtained introducing a restriction in the morphology of λ-formulae. We will not allow M to be a λ-formula, but only a variable or a constant; we will obtain this modification replacing, according to the Backus notation that defines the language of the λ-calculus, the equation

$$\langle\text{formula}\rangle = (\langle\text{formula}\rangle\ \langle\text{formula}\rangle)$$

with the following equation

$$\langle\text{formula}\rangle = (\langle\text{formula}\rangle\ \langle\text{variable}\rangle) \mid (\langle\text{formula}\rangle\ \langle\text{constant}\rangle) .$$

4. Machines Implementing T_2 Mappings

Now we want to discuss the problem of defining a transducer (i.e. an abstract automaton with output) capable of implementing T_2 mappings.

As it is well known, the general case of the automatic reduction to normal form of the λ-formulae can not be treated by means of a machine less powerful than a Turing machine; see Böhm and Dezani [2] for a description of such a machine. Since class T_2 implies a form of substitution less general than the one of the λ-calculus, that is, since T_2 has a power equivalent to the power of the subclass, above considered, we will surely succeed in implementing T_2-mappings by means of a less powerful machine. We will find our machine in the classical hierarchy of automata at the level of stack automata. As we will see later, a stack is certainly necessary to store the symbol table, which is constructed and used during the computation.

Our construction will show that there exists a machine M, which is a ***universal machine for class T_2***. This machine is a two-way deterministic stack automaton.

Two-way deterministic stack automata were first introduced by Ginsburg and others [6]. However, here we will closely follow the Hopcroft and Ullman [7] definition. A ***two-way deterministic stack automaton*** is an abstract machine provided with two lists. The first list is read, but not changed, by a "pointer" which travels up and down the list as often as desired. This list is referred to as the "input list" or simply the "input".

A second list, called "stack", is also used. This stack is a last-in, first-out store, i.e., it may be written or erased from the top of the stack in the conventional way. Moreover, the internal part of the stack may be read, but not rewritten; so the stack contents may be changed only from the top.

An output structure is added to the machine in order to produce strings. The machine may be in one out of a finite set of internal states. To work, the machine must be in a certain state reading both an input and a stack symbol. The device then acts as follows:

(i) it goes to another state;

(ii) it moves, at most, one symbol to the left or to the right of the input symbol just read;

(iii) it does exactly one of two alternatives: (a) it moves its stack pointer, at most, one symbol to the left or to the right of the stack symbol just read; (b) if the pointer is scanning the rightmost stack symbol, it may write a (possibly empty) finite sequence of symbols on the stack, erasing simultaneously the symbol just read.

(iv) it outputs a finite sequence of symbols.

We will now formalize the model.

Definition 4.1. A *stack transducer* is a 8-tuple $S = (Q, \Sigma, \Gamma, Y, \delta, \delta_b, q_0, Z_0)$ where

1) $Q = \{q_0, q_1, \ldots\}$ is a finite non-empty set of *states*.
2) $\Sigma = \{x_1, x_2, \ldots\}$ is a finite non-empty set of *inputs*.
   Σ includes the "left and right end markers", $\mathcal{C}$ and \$, respectively, which are assumed to be the first and last symbols of any input.
3) Γ is a finite non-empty set of *stack symbols*.
4) Y is a finite non-empty set of *outputs*,
5) Z_0 is the *start symbol* and acts as a marker of left end of the stack.
6) q_0 is the *start state*.
7) δ is a mapping from $Q \times \Sigma \times \Gamma$ into subsets of
$$\{-1, 0, 1\} \times Q \times \{-1, 0, 1\} \times Y^*$$
8) δ_b is a mapping from $Q \times \Sigma \times \Gamma$ into subsets of
$$\{-1, 0, 1\} \times Q \times (\{-1, 0, E\} \cup \Gamma^*) \times Y^*$$

δ is the *next move mapping* for the condition when the stack pointer is not scanning a blank. Under this condition, the stack transducer may change state and move its two heads and produce an output.

The significance of $\delta(q, a, Z) = (D_1, p, D_2, w)$ is that, if S is in state q scanning "a" on its input tape and Z on its stack, it may go to state p and move its input and stack pointers in directions D_1 and D_2, respectively. That is, if D_1 or D_2 has value -1, 1 or 0, the relevant pointer will move to the left, to the right, or remain stationary, respectively. Moreover, the (possibly empty) string w is produced as output.

δ_b is the *next move mapping* for the condition when the stack pointer is scanning a *blank*. It will always be the case that this blank is the leftmost blank on the stack. The leftmost blank will be called the *top of stack*. We suppose that the next move can depend upon the rightmost nonblank symbol even though the stack pointer is not scanning it currently.

The significance of $\delta_b(q, a, Z) = (D, p, X, w)$ is that if S is in state q scanning "a" on its input tape and a blank on its stack, with the nonblank symbol Z immediately to the left, then S may enter state p, move its input pointer in direction D, output the string w, and make the following motion of its stack pointer.

If $X = 0$, no move of the stack pointer.
If $X = -1$, the stack pointer moves to the left.
If $X = E$, the stack pointer moves to the left and prints a blank on the new cell scanned (i.e., the rightmost nonblank is erased). The cell now scanned becomes the new top of stack.
If X is a string of Γ^*, the stack pointer prints X over the stack and then moves to the right, to the new top of stack. We assume that X is not Z_0.

Note that the stack contents are always Z_0, followed by a nonblank string to the right and an infinite sequence of blanks to the right of that. Another useful concept we need to introduce is that of Chomsky parsed string.

Given a context-free grammar G, with n production rules, we shall consider grammar $\overline{G}$ defined on the alphabet $N \cup Z \cup \bigcup_{i=l}^{n} \{[_i,]_i,\}$ in the following way:

If $A \to \alpha$ is the i-th production rule of grammar $G = (N, Z, P, S)$, $A \to [_i\alpha]_i$ is the i-th rule of grammar $\overline{G}$.

That is, each production rule of $\overline{G}$ is obtained enclosing the corresponding production of G by a couple of characters $[_i,]_i$. Grammars of type $\overline{G}$ are called *parenthesized grammars*. To each string w of the language $L(G)$ we may associate that string $\overline{w}$ of $L(\overline{G})$ which is generated using the productions corresponding to those that were used to generate w. The string $\overline{w}$ is unique iff G is unambigous. Se shall call $\overline{w}$ the *Chomsky parsed string*.

The T_1-semantics will be defined by giving the n-tuple of representing words. The more general representing word w_i, associated to the i-th rule of G, is obtained by taking any string σ_i in the alphabet $\{x_1, \ldots, x_M\}$, where M is max(length (α)), say $\sigma_i = x_4x_1x_1x_2x_1x_3$, and producing the word $w_i = w_{i,1}x_4w_{i,2}x_1w_{i,3}x_1w_{i,4}x_2w_{i,5}x_1w_{i,6}x_3w_{i,7}$ where $w_{i,h} \in Y^*$. Instead of actually writing the value of the strings $w_{i,h}$, we can embed the matrix $w_{i,h}$ into the finite part of the automaton and consider, instead of w_i, the word $v_i = c_6x_3c_5x_1c_4x_2c_3x_1c_2x_1c_1x_4 \mid_i$ in the alphabet $\{\mid_1, \mid_2, \ldots, \mid_n, x_1, x_2, \ldots, x_M, c_1, c_2, \ldots, c_K\}$ where $K = \max(\text{length } (\sigma_i))$ for $i = l, \ldots, n$. v_i is obtained as a mirror image of w_i substituting the strings $w_{i,k}$ with the symbols c_h and adding the symbols $\mid_i$, telling us which of the n-production rules is being considered. The word w_i has been reversed because it will be printed on the stack and successively accessed by the machine from right to left.

Informal Machine Description

1) Whenever the input reading head encounters an open nonterminal parenthesis, i.e. a symbol $[_i$, the corresponding semantical rule v_i is printed on the stack, initially set to Z_0. Actually within rule v_i the symbol $\mid_i$ is shifted two positions towards the left. Thus, for the above example, v_i is written as v_i':

$$v_i' = c_6a_3c_5a_1c_4a_2c_3a_1c_2a_1 \mid_i c_1a_4$$

where we have used the letters $a_1, a_2, \ldots$ instead of $x_1, x_2 \ldots$

2) The rightmost symbol of v_i', a_4 in the above example, tells machine M that the input head has to reach the fourth nest of parentheses to the right. To control the movement of the input reading head, an additional symbol is written on the stack. This symbol is initially a_1. Each time a complete nest of parentheses is skipped, the additional symbol a_i is changed into a_{i+1}. The proper nest is reached only when the two last symbols on the tape are equal. When the wanted nest is reached — in our example this, initially, is the fourth nest — then M leaves this configuration on the stack

$$\ldots c_6a_3c_5a_1c_4a_2c_3a_1c_2a_1 \mid_i a_4Z_0 \tag{1}$$

and cycles back to comma 1) so as to process the fourth nest.

3) When the fourth nest is completely processed, the stack will be again as in (1) and the input reading head will be positioned just after the fourth nest. At this point, the input head has to skip 4 nests back, i.e. to the left, in order to be positioned exactly at the beginning of the substring enclosed in the parentheses of the i-th rule. M will perform this back-movement using the same technique of counting each skipped nest. At the end, the stack configuration will be

$$\ldots Z_0c_6a_3c_5a_1c_4a_2c_3a_1c_2a_1 \mid_i .$$

Successively, the machine will look for the symbols preceding the $|_i$ barrier. If there is a Z_0, then this means that the nest, so far analyzed as the i-th rule, has already been completely processed and M will cycle back to comma 3). If there are still any symbols of type a_i or c_j, then the first couple of symbols to the right will jump the $|_i$ barrier. In our case we would have the following configuration on the stack:

$$\cdots Z_0 c_6 a_3 c_5 a_1 c_4 a_2 c_3 a_1 \,|_i\, c_2 a_1$$

and the machine would then cycle back to comma 2).

1′) Whenever the input head encounters the symbol $[_\lambda$, (which encloses the string on the right-hand side of a production, to which the λ-operator is associated), the contents of the following two couples of parentheses ($[_t\ ]_t$, $[_{id}\ ]_{id}$) are printed on the stack, using suitable separation markers. A symbol table is so constructed on the stack to contain the couples (type, identifier). Then the machine cycles back to 1), 1′) or 1″).

1″) Whenever the input reading head encounters the symbol $[_{id}$ and the internal state is [+, +, start], the machine compares the string contained in $[_{id}\ ]_{id}$ with the strings of the symbol table. If such a string is found, the string associated to it in the symbol table is immediately output. Then M erases the symbol table and cycles back to 3).

The output of M is a function of its internal state and is produced under the following circumstances:

(a) A terminal t is found on the input tape. Then if the internal state is [+, +, start], it produces output $\mu(t)$;

(b) when symbol c_k is read on the stack and M is in state $[i, x, eq_1]$, the output is $w_{i,k}$;

(c) when symbol Z_0 is read on the stack and M is in the state $[i, +, lend_2]$, the output is $w_{i,\text{last}}$ (the last $w_{i,h}$ of the i-th rule);

(d) when a string equal to the one in $[_{id}\ ]_{id}$ is found on the symbol table and M is in the state COPY, the output is the left-hand element of the couple, whose right-hand element is the content of $[_{id}\ ]_{id}$.

Identity operators will always be associated to the parentheses $[_{id}\ ]_{id}$ and $[_t\ ]_t$, therefore during translation the contents of these parentheses will not be modified. However from the point of view of the machine, these operators are different from ordinary identity operators. That is, when the input reading head encounters the symbol $[_{id}$ or the symbol $[_t$, the machine does not write the rule on the stack, but, according to its internal state, it writes the symbol table on the stack or compares and then substitutes the contents of the parentheses, as the symbol table shows.

Formal Machine Description

The following formulae define a stack deterministic two-way stack machine M that performs the mapping $\bar{\alpha} \rightarrow \mu(\alpha)$, for a given μ of class T_2. Letter i denotes a variable ranging over the set of the first N integers. Moreover, the "don't care" symbol "—" will be used to denote any symbol that can possibly occur in that symbol position.

[+, +, start], Z_0, $\mathscr{E} \overset{\delta_b}{\rightarrow}$ (1, [+, +, start], 0, $\varnothing$). (Skip over $\mathscr{E}$)

[+, +, start], Z_0, $[_i \overset{\delta_b}{\rightarrow}$ (1, $[i, +, eq]$, i-rule, $\varnothing$). (M writes the rw)

[+, +, start], Z_0, $[_\lambda \overset{\delta_b}{\rightarrow}$ (1, [+, +, start], 0, $\varnothing$). (M starts reading and writing the first two arguments on the stack)

$[+, +, \text{start}], Z_0, [_\iota \xrightarrow{\delta_b} (1, \text{ST}, 0, \varnothing)$. (*M* starts writing the symbol-table)

$[+, +, \text{start}], Z_0, [_{\text{id}} \xrightarrow{\delta_b} (1, \text{CFR}, 0, \varnothing)$. (*M* starts comparing)

$[+, +, \text{start}], -, T \xrightarrow{\delta_b} (0, [+, +, b\text{-nest}_1], 0, \mu(T))$ for any terminal symbol T

. . . .

$[i, +, \text{eq}], X, - \xrightarrow{\delta_b} (0, [i, X, \text{eq}_1], E, \varnothing)$. $\quad X \in \{a_i\}$

$[i, +, \text{eq}], Z_0, - \xrightarrow{\delta_b} (-1, [+, a_1, \text{lend}_1], E, \varnothing)$

M starts comparing the last two a_i on the stack.

. . . .

$[i, X, \text{eq}_1], Y, - \xrightarrow{\delta_b} (0, [i, X, \text{eq}_1], E, \varnothing) \quad Y \in \{a_i\}, \quad X = Y$

$[i, X, \text{eq}_1], F, - \xrightarrow{\delta_b} (0, [+, +, \text{start}], E$ and print $XZ_0, w_{i,k})$ if $F = c_k \in \{c_r\}$

$[i, X, \text{eq}_1]\ Y, - \xrightarrow{\delta_b} (0, [i, +, -f\text{-nest}], S(X), \varnothing)$ where $Y \in \{a_i\}$, $X \neq Y$ and $S(a_i) = a_{i+1}$

M compares X with the current symbol on the top of the stack.

. . . .

$[i, +, f\text{-nest}], -, Z \xrightarrow{\delta_b} (1, [i, +, f\text{-nest}_1], C, \varnothing) \qquad Z \in \{[_j\}$

$[i, +, f\text{-nest}], Y, T \xrightarrow{\delta_b} (1, [i, Y, \text{eq}_1], E, \varnothing)$ T is a terminal, $\quad Y \in \{a_i\}$

$[i, +, f\text{-nest}_1], C, Z \xrightarrow{\delta_b} (1, [i, +, f\text{-nest}], C, \varnothing) \qquad Z \in \{[_j\}$

$[i, +, f\text{-nest}_1], C, T \xrightarrow{\delta_b} (1, [i, +, f\text{-nest}_1], 0, \varnothing)$

$[i, +, f\text{-nest}_1], C, W \xrightarrow{\delta_b} (1, [i, +, f\text{-nest}_1], E, \varnothing) \qquad W \in \{[_j\}$

$[i, +, f\text{-nest}_1], Y, - \xrightarrow{\delta_b} (0, [i, Y, \text{eq}_1], E, \varnothing) \qquad Y \in \{a_i\}$

M skips a nest

. . . .

$[+, +, b\text{-nest}], -, - \xrightarrow{\delta_b} (-1, [+, +, b\text{-nest}_1], 0, \varnothing)$

$[+, +, b\text{-nest}_1], H, T \xrightarrow{\delta_b} (0, [+, S(H), \text{lend}_1], E, \varnothing)$

$[+, +, b\text{-nest}_1], -, W \xrightarrow{\delta_b} (-1, [+, +, b\text{-nest}_2], D, \varnothing) \qquad W \in \{]_j\}$

$[+, +, b\text{-nest}_2], D, W \xrightarrow{\delta_b} (-1, [+, +, b\text{-nest}_2], D, \varnothing)$

$[+, +, b\text{-nest}_2], D, T \xrightarrow{\delta_b} (-1, [+, +, b\text{-nest}_2], 0, \varnothing)$

$[+, +, b\text{-nest}_2], D, Z \xrightarrow{\delta_b} (-1, [+, +, b\text{-nest}_2], E, \varnothing) \qquad Z \in \{[_j\}$

$[+, +, b\text{-nest}_2], H, - \xrightarrow{\delta_b} (+1, [+, S(H), \text{lend}_1], E, \varnothing)$ where $H \in \{Z_0, a_1, a_2, \ldots, a_n\}$
and S is a function defined by $S(Z_0) = a_1$, $S(a_i) = a_{i+1}$, for $i < n$

M skips a nest, moving back

. . . .

$[+, X, \text{lend}_1], Y, - \xrightarrow{\delta_b} (0, [+, +, \text{lend}_2], E, \varnothing)$ $\quad X = Y, \quad Y \in \{a_i\}$

M compares X with Y (Y indicates the actual nesting level)

$[+, X, \text{lend}_1], Y, - \xrightarrow{\delta_b} (0, [+, +, b\text{-nest}], X, \varnothing)$ $\quad X \neq Y$

. . . .

$[+, +, \text{lend}_2], |_i, - \xrightarrow{\delta_b} (0, [i, +, \text{lend}_2], E, \varnothing)$

as $X = Y$, the machine erases the symbol $|_i$ from the stack

$[i, +, \text{lend}_2], Z_0, - \xrightarrow{\delta_b} (-1, [+, a_1, \text{lend}_1], E, w_{i, \text{last}})$

$[i, +, \text{lend}_2], X, - \xrightarrow{\delta_b} (0, [i, X, \text{lend}_3], E, \varnothing)$

$[i, X, \text{lend}_3], F, - \xrightarrow{\delta_b} (0, [i, +, \text{eq}], E$ and print $|_i FXa_1, \varnothing)$

. . . .

$ST, -, T \xrightarrow{\delta} (1, ST, T, \varnothing)$

$ST, -,]_i \xrightarrow{\delta} (1, ST, \#, \varnothing)$

$ST, -, [_{\text{id}} \xrightarrow{\delta} (1, ST, 0, \varnothing)$

$ST, -,]_{\text{id}} \xrightarrow{\delta} (1, [+, +, \text{start}], Z_1, \varnothing)$

M writes the symbol-table on the stack

. . . .

$\text{CFR}, Z_0, T \xrightarrow{\delta_b} (1, \text{CFR}, 0, \varnothing)$

$\text{CFR}, Z_0,]_{\text{id}} \xrightarrow{\delta_b} (0, \text{CFR}, Z_2, \varnothing)$

$\text{CFR}, \female,]_{\text{id}} \xrightarrow{\delta} (0, \text{CFR}, -1, \varnothing)$ $\female$ = any symbol different from Z_1, Z_2

$\text{CFR}, Z_1,]_{\text{id}} \xrightarrow{\delta} (-1, \text{CFRS}, -1, \varnothing)$

$\text{CFRS}, T_1, T_2 \xrightarrow{\delta} (-1, \text{CFRS}, -1, \varnothing)$ $\quad T_1 = T_2$ (terminals)

$\text{CFRS}, T_1, T_2 \xrightarrow{\delta} (+1, \text{CFR}, -1, \varnothing)$ $\quad T_1 \neq T_2$ (terminals)

$\text{CFRS}, \#, [_{\text{id}} \xrightarrow{\delta} (0, \text{COPY}, -1, \varnothing)$

$\text{CFR}, \female, T \xrightarrow{\delta} (1, \text{CFR}, 0, \varnothing)$

M compares symbol table item with input item

. . . .

$\text{COPY}, T, [_{\text{id}} \xrightarrow{\delta} (0, \text{COPY}, -1, T)$

$\text{COPY}, Z_j, [_{\text{id}} \xrightarrow{\delta} (0, \text{TOP}, 1, \varnothing)$ $\quad j = 0, 1$

the type is produced as output

$\text{TOP}, +, [_{\text{id}} \overset{\delta}{\rightarrow} (0, \text{TOP}, 1, \varnothing)$ $\quad + =$ any symbol different from Z_0, Z_2

$\text{TOP}, Z_2, [_{\text{id}} \overset{\delta_b}{\rightarrow} (0, \text{TOP}, E, \varnothing)$

$\text{TOP}, Z_0, [_{\text{id}} \overset{\delta_b}{\rightarrow} (0, [+, X, \text{lend}_1], E, \varnothing)$

. . . .

$[i, +, \text{lend}_2], Z_1, [_j \overset{\delta_b}{\rightarrow} (-1, \text{CLEAR}, E, \varnothing)$

$\text{CLEAR}, *, - \overset{\delta_b}{\rightarrow} (-1, \text{CLEAR}, E, \varnothing)$ $\quad * =$ any symbol different from Z_0, Z_1

$\text{CLEAR}, Z_j, - \overset{\delta_b}{\rightarrow} (-1, \text{CLEAR}, 0, \varnothing)$ $\quad j = 0, 1$

$\text{CLEAR}, Z_0, [_\lambda \overset{\delta_b}{\rightarrow} (0, [+, a_1, \text{lend}_1], E, \varnothing)$

$\text{CLEAR}, Z_1, [_\lambda \overset{\delta_b}{\rightarrow} (-1, \text{CLEAR}, E, \varnothing)$

having performed the substitutions of a certain nest, M erases the symbol table of that nest.

We have the following

Theorem. *Let S be a T_2-semantics, defined by a set of representing words and λ-operators on an unambigous grammar G. Let μ be the mapping of the language $L(G)$ defined by S and the set of initial conditions $\mu(t) = t'$ for $t \in T$ (terminal alphabet of G). The machine M above defined, where the i-rule is derived from the corresponding representing word in the manner shown in comma 1) of the transducer informal description, transforms any parsed string $\bar{\alpha}$ of the parsed language $\bar{L}$ into $\mu(\alpha)$.*

The proof is omitted here; but it can easily be worked out from the description of M.

Example 1. Let G be following grammar, where an operator of class T_2 is associated to each production:

$S \rightarrow TIS$	$\lambda(x_1, x_2, x_3)$
$S \rightarrow TIB$	$\lambda(x_1, x_2, x_3)$
$B \rightarrow bEI$	$O_1(x_1, x_2, x_3) = w_{1,1}x_3w_{1,2}x_3w_{1,3}$
$T \rightarrow p$	$O_t(x_1) \equiv x_1$
$T \rightarrow r$	$O_t(x_1) \equiv x_1$
$E \rightarrow x$	$O_2(x_1) \equiv x_1$
$I \rightarrow y$	$O_{\text{id}}(x_1) \equiv x_1$.

The input string is $\alpha = pybxy$.

It is easy to see that this string has substantially the same structure as the ALGOL-like program

integer y; ... y ... (statement containing y).

The input Chomsky parsed string is:

$$\bar{\alpha} = [_\lambda[_t p]_t\, [_{\text{id}} y]_{\text{id}}\, [_1 b [_2 x]_2\, [_{\text{id}} y]_{\text{id}}]_1]_\lambda \,.$$

Machine S is defined by:

$$\Sigma = \{p, r, x, y, [_\lambda,]_\lambda, [_t,]_t, [_{\text{id}},]_{\text{id}}, [_1,]_1, [_2,]_2\}$$

$$\Gamma = \{|_1, |_2, C, Z_0, Z_1, Z_2, a_1, a_2, a_3, c_1, c_2, c_3, D, p, r, x, y, \#\}$$

Q is the union of the following two sets of states:

1) a subset of the cartesian product of three sets: s_1, s_2, s_3, where s_1 is set $\{+, 1, 2, 3\}$, s_2 is set $\{+, a_1, a_2, a_3\}$ and s_3 is set {start, eq, eq$_1$, f-nest, f-nest$_1$, b-nest, b-nest$_1$, b-nest$_2$, lend, lend$_1$, lend$_2$, lend$_3$}.

2) {ST, CFR, CFRS, COPY, TOP}
   The semantics rule of operator O_1 is:

$$c_2a_3 \mid {}_1c_1a_3 .$$

   The semicolon on the input tape and on the stack represents the position of the reading head.

We have that $\mu(pybxy) = w_{1,1}pw_{1,2}pw_{1,3}$.

Computation of $\mu(pybxy)$

Machine State	Stack	Input Tape	
[+, +, start]	$Z_0;$	$;\ \mathcal{E}[_\lambda[_t p]_t\ [_{id} \dots$	
[+, +, start]	$Z_0;$	$\mathcal{E};\ [_\lambda[_t p]_t \dots$	
[+, +, start]	$Z_0;$	$\mathcal{E}[_\lambda;\ [_t p]_t \dots$	
ST	$Z_0;$	$\mathcal{E}[_\lambda[_t;\ p_t] \dots$	(a)
ST	$Z_0p;$	$\mathcal{E}[_\lambda[_t p;]_t \dots$	
ST	$Z_0p\ \#$	$\mathcal{E}[_\lambda[_t p]_t;\ [_{id} \dots$	
ST	$Z_0p\ \#;$	$\dots p]_t\ [_{id};\ y \dots$	
ST	$Z_0p\ \#\ y;$	$\dots p]_t\ [_{id}y;]_{id} \dots$	
[+, +, start]	$Z_0p\ \#\ yZ_1;$	$\dots [_{id}y]_{id};\ [_1 \dots$	
[1, +, eq]	$\dots yZ_1c_2a_3 \mid {}_1c_1a_3a_1;$	$\dots [_{id}y]_{id}\ [_1;\ b \dots$	
[1, a_1, eq$_1$]	$\dots a_3 \mid_1 c_1c_3;$	$\dots [_{id}y]_{id}\ [_1;\ b \dots$	
[1, +, f-nest]	$\dots a_3 \mid_1 c_1a_3a_2;$	$\dots [_{id}y]_{id}\ [_1;\ b \dots$	
[1, a_2, eq$_1$]	$\dots a_3 \mid_1 c_1a_3;$	$\dots [_{id}y]_{id}\ [_1b;\ [x_2 \dots$	
[1, +, f-nest]	$\dots a_3 \mid_1 c_1a_3a_3;$	$\dots [_{id}y]_{id}\ [_1b;\ [_2x \dots$	
[1, +, f-nest$_1$]	$\dots a_3 \mid_1 c_1a_3a_3C;$	$\dots y]_{id}\ [_1b[_2;\ x \dots$	(b)
[1, +, f-nest$_1$]	$\dots a_3 \mid_1 c_1a_3a_3C;$	$\dots y]_{id}\ [_1b[_2x]_2 \dots$	
[1, +; f-nest$_1$]	$\dots a_3 \mid_1 c_1a_3a_3;$	$\dots y]_{id}\ [_1b[_2x]_2;\ [_{id}$	
[1, a_3, eq$_1$]	$\dots a_3 \mid_1 c_1a_3;$	$\dots y]_{id}\ [_1b[_2x]_2;\ [_{id}$	
[1, a_3, eq$_1$]	$\dots a_3 \mid_1 c_1;$	$\dots y]_{id}\ [_1b[_2x]_2;\ [_{id}$	

Output: $w_{1,1}$

[+, +, start]	$\dots yZ_1c_2a_3 \mid_1 a_3Z_0;$	$\dots [_1b[_2x]_2;\ [_{id}y]_{id}$	
CFR	$\dots \mid_1 a_3Z_0;$	$\dots [_{id};\ y]_{id}$	
CFR	$\dots a_3 \mid_1 a_3;\ Z_0$	$\dots [_{id}y]_{id}$	

. . . .

. . . (M moves the stack pointer to the left until Z_1 is encountered) . . . .

. . . .

CFR	$\dots y;\ Z_1c_2a_3 \mid_1 \dots$	$\dots [_{id}y;\ ]_{id}$	
CFRS	$\dots \#;\ yZ_1c_2 \dots$	$\dots [_{id};\ y]_{id}$	
COPY	$Z_0;\ p\ \#\ yZ_1 \dots$	$\dots;\ [_{id}y]_{id}$	(c)

Output: p

COPY	$;\ Z_0p\ \#\ yZ_1 \dots$	$\dots;\ [_{id}y]_{id}$	
TOP	$Z_0;\ p\ \#\ y \dots$	$\dots;\ [_{id}y]_{id}$	

. . . (M moves the pointer back to the top) . . . .

TOP	$\dots \mid_1 a_3Z_0;$	$\dots;\ [y_{id}]_{id}$	

$[+, a_1, \text{lend}_1]$	$\dots \|_1\, a_3;$	$\dots ;\, [_{\text{id}} y]_{\text{id}}$	
$[+, +, b\text{-nest}]$	$\dots \|_1\, a_3 a_1;$	$\dots ;\, [_{\text{id}} y]_{\text{id}}$	
$[+, +, b\text{-nest}_1]$	$\dots \|_1\, a_3 a_1;$	$\dots [_2 x;\,]_2\, [_{\text{id}} y]_{\text{id}}$	
$[+, +, b\text{-nest}_2]$	$\dots \|_1\, a_3 a_1 D;$	$\dots [_2;\, x]_2\, [_{\text{id}} y]_{\text{id}}$	
$[+, +, b\text{-nest}_2]$	$\dots \|_1\, a_3 a_1 D;$	$\dots [_1 b;\, [_2 x]_2\, [_{\text{id}} \dots$	
$[+, +, b\text{-nest}_2]$	$\dots \|_1\, a_3 a_1;$	$\dots [_1;\, b[_2 x]_2\, [_{\text{id}} \dots$	
$[+, a_2, \text{lend}_1]$	$\dots \|_1\, a_3;$	$\dots [_1 b;\, [_2 x]_2\, [_{\text{id}} \dots$	(d)
$[+, +, b\text{-nest}]$	$\dots \|_1\, a_3 a_2;$	$\dots [_1 b;\, [_2 x]_2\, [_{\text{id}} \dots$	
$[+, +, b\text{-nest}_1]$	$\dots \|_1\, a_3 a_2;$	$\dots [_1;\, b[_2 x]_2\, [_{\text{id}} \dots$	
$[+, a_3, \text{lend}_1]$	$\dots \|_1\, a_3;$	$\dots [_1;\, b[_2 x]_2\, [_{\text{id}} \dots$	
$[+, +, \text{lend}_2]$	$\dots Z_1 c_2 a_3 \|_1;$	$\dots [_1;\, b[_2 x]_2\, [_{\text{id}} \dots$	
$[1, +, \text{lend}_2]$	$\dots Z_1 c_2 a_3;$	$\dots [_1;\, b[_2 x]_2\, [_{\text{id}} \dots$	
$[1, a_3, \text{lend}_3]$	$\dots Z_1 c_2;$	$\dots [_1;\, b[_2 x]_2\, [_{\text{id}} \dots$	
$[1, +, \text{eq}]$	$\dots Z_1 \|_1\, c_2 a_3 a_1$	$\dots [_1;\, b[_2 x]_2\, [_{\text{id}} \dots$	

The computation goes forward, running through the steps of block (b) again and, then, through blocks (c) and (d). At the end, having output the last constant, the machine erases the symbol table and stops.

Let us now see some other meaningful examples of computations of the stack transducer, which show how the transducer strictly respects the ALGOL rules given by the scope definition of variables.

Example 2. Let $\bar{\alpha} = [_1[_2 w]_2\, [_\lambda[_\iota a]_\iota\, [_{\text{id}} b]_{\text{id}}\, [_2 w']_2]_\lambda\, [_3 w'']_3]_1$ be the input parsed string, where w, w', and w'' are any Chomsky parsed strings. Furthermore $O_1 = x_2 x_3$.

We shall now list the main configurations that the stack and the input tape assume during the computation.

STACK	INPUT TAPE
$Z_0 c_2 a_3 \|_1\, c_1 a_2 a_1$	$;[_1[_2 w]_2 \dots$

After a certain number of moves, described in comma 2) of the transducer informal description, M eventually reaches the following configuration:

$Z_0 c_2 a_3 \|_1\, a_2 Z_0$	$[_1[_2 w]_2;\, [_\lambda[_\iota a \dots$

After a certain number of moves, M writes the symbol table on the stack:

$\dots a_3 \|_1\, a_2 Z_0 a \# b Z_1$	$\dots [_{\text{id}} b]_{\text{id}};\, [_2 w']_2 \dots$

Then, M "translates" string w', substituting "a" for each occurrence of "b" in w'. Then M erases the symbol table and moves the pointer to the beginning of the second nest:

$\dots a_3 \|_1\, a_2$	$\dots [_1[_2 w]_2;\, [_\lambda \dots$

M will continue the computation, as we have already seen in detail in the previous example.

Example 3. Let $\bar{\alpha} = [_\lambda[_\iota a]_\iota\, [_{\text{id}} b]_{\text{id}}\, [_\lambda[_\iota c]_\iota\, [_{\text{id}} d]_{\text{id}}\, [_1 w]_1]_\lambda]_\lambda$ be the input string, where w is any Chomsky parsed string.

STACK	INPUT TAPE
$Z_0 a \# b Z_1$	$[_\lambda[_\iota a]_\iota\, [_{\text{id}} b]_{\text{id}};\, [_\lambda \dots$

Starting from this configuration, the machine follows the same sequence of states it followed till this point and eventually reaches the configuration:

$Z_0 a \# b Z_1 c \# d Z_1$	$\dots [_\lambda[_\iota c]_\iota\, [_{\text{id}} d]_{\text{id}};\, [_1 w]_1\,]_\lambda]_\lambda\,.$

Then M computes string w: when it encounters the symbol $[_{\text{id}}$, it moves the stack pointer back till it encounters Z_1 and carries out the comparison, as we showed in detail in Example 1. If the comparison is not positive, M moves the stack pointer back again till it encounters another Z_1 and so on.

Example 4. Let $\bar{\alpha} = [_\lambda[_\iota a]_\iota\, [_{\mathrm{id}} b]_{\mathrm{id}}\, [_1[_\lambda[_\iota c]_\iota\, [_{\mathrm{id}} d]_{\mathrm{id}}\, [_2 \cdots [_{\mathrm{id}} b]_{\mathrm{id}} \cdots [_{\mathrm{id}} d]_{\mathrm{id}}]_2]_\lambda\, [_3 \cdots [_{\mathrm{id}} b]_{\mathrm{id}} \cdots$ $\cdots]_3]_1]_\lambda$ be the input string.

After a certain number of moves, the transducer will reach the following configuration:

STACK	INPUT TAPE
$Z_0 a \# b Z_1$	$[_\lambda[_\iota a]_\iota\, [_{\mathrm{id}} b_{\mathrm{id}}];\ [_1 \cdots$

Now, behaving in the way described at comma 1) of the informal description, M writes on the stack the rw associated to the operator $O_1 \equiv x_1 x_2$:

$Z_0 a \# b Z_1 c_2 a_2 \mid_1 c_1 a_1 a_1$	$\cdots [_1;\ [_\lambda[_\iota c]_\iota \cdots$

As we have seen in the other examples, M will reach the configuration:

$Z_0 a \# b Z_1 c_2 a_2 \mid_1 a_1 Z_0 c \# d Z_1$	$\cdots [_{\mathrm{id}} d]_{\mathrm{id}};\ [_2 \cdots$

As it did in Example 3, M will compute the contents of the parentheses $[_2,]_2$. Obviously, during the searches made by the machine for couples on the symbol table, after the pointer has tested the first couple, it moves in search of the first Z_1, thus jumping the rw symbols.

At the end of this computation the configuration is:

$Z_0 a \# b Z_1 c_2 a_2 \mid_1 a_1$	$\cdots [_1;\ [_\lambda[_\iota c]_\iota \cdots$

Then, M will compute the contents of $[_3]_3$.

5. An Extension of GSDT Schemes and their Implementation

5.1. Terminology

Definition 5.1. A *generalized syntax directed translation scheme* (GSDTS) is a generalization of a context-free grammar. It will be denoted by a 6-tuple $V = (N, T, T', \Gamma, R, S)$ where N, T, T' are finite sets of *nonterminals*, *input symbols* and *output symbols*, respectively.

Γ is a finite set of *translation symbols* of the form A_i, where A_i is in N and i is an integer, plus symbol S_1. S, in N, is the *start symbol*. R is a set of rules of the form

$$A \to \alpha\,, \quad A_1 = \beta_1\,, \quad A_2 = \beta_2, \ldots, A_m = \beta_m$$

subject to the following conditions:

1) $A_j \in \Gamma, \quad 1 \leqq j \leqq m$.
2) Each β_i either belongs to T' or is a symbol B_k of Γ, such that B is a nonterminal occurring in α.
3) If α contains several symbols B, then each B_k in β_i is associated by means of an apex, with one of these instances of B.

We call $A \to \alpha$ the *base production* of the rule; we call A_i a *translation* of A, and $A_i = \beta_i$ a *translation element* associated to this rule. If P denotes the set of the base productions of all the rules, then $G = (N, T, P, S)$ is the base grammar of V.

We shall now define the output of a GSDTS. A string for every A_i is associated with each interior node of the parse tree labeled A. This string will be called *value* (or translation) of A_i at the node n. Each value is computed by substituting the values, defined at the direct descendants of n, for each translation symbol of the translation element A_i. The right translation element for A_i is the one associated with the production used at n. For example suppose that

$$A \to BaC\,, \quad A_1 = bB_1C_2B_1\,, \quad A_2 = C_1cB_2$$

is a rule of a GSDTS and the base production $A \to BaC$ is used to expand a node labeled A in the parse tree, as is shown in Fig. 2. Then, if the values of B_1 and B_2 at node B, and of C_1 and C_2 at node C are those of Fig. 2, the value of A_1, defined at the node labeled A, is $bx_1y_2x_1$ and the value of x_2 at the same node is y_1cx_2.

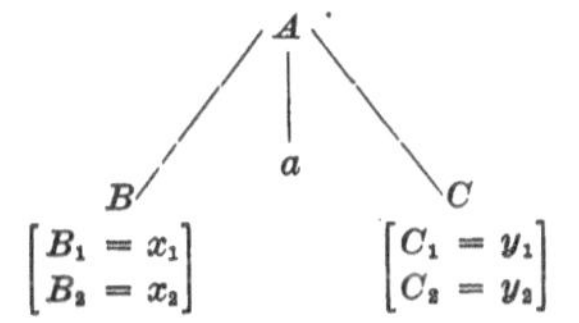

Fig. 2

The *translation defined by* V, denoted by $\tau(V)$, is the set $\{(x, y) \mid x$ has a parse tree in the base grammar of V and y is the value of S_1 at the root of this tree$\}$.

The GSDT scheme we have just defined, is an extension of the T_1-semantics. The latter are GSDT's with only one translation element defined at each node. If we restrict the GSDT further so as to require that the formula for the translation element has exactly one occurrence of each nonterminal of the base production, then we have the SDT.

5.2. The pushdown transducer implementing GSDT's

The stack transducer we introduced in the previous section is a generalization of the *pushdown transducer* of Petrone [11]. The stack transducer (st) is obtained from the pushdown transducer (pdt), allowing the pointer to read inside the pushdown tape and introducing some new internal states and definition formulae, as one can see, comparing the definitions of the two transducers.

The pdt is capable of implementing T_1-mappings; we will now show that, making slight changes on the set of formulae defining the pdt, it is possible to make the pdt capable of implementing GSDT's. We know that GSDT's differ from T_1-semantics in having several translation elements associated to each production, that corresponds to associate several representing words to each production.

Now the problem is to find an *rw*, which, besides showing which nonterminal is concerned in the translation, and how many times and in which order its translation must occur, it also shows which of the translations associated to this terminal must be used. Let us make the problem clear with an example.

Suppose we have the following simple GSDTS:

1) $\quad S \to bS \qquad S_1 = w_{1,1}S_2w_{1,2}S_1w_{1,3}$

2) $\quad S \to a \qquad S_1 = a \quad S_2 = d\,.$

Now we want to find a representation of translation elements, to be written on the pushdown list, easily understandable by the machine. Let us follow the pdt computation step by step and decide where and how to make changes.

Let $\bar{\alpha} = [_1b[_2a]_2]_1$ be the input parsed string, the translation of which is $\mu(\alpha) = \mu(ba) = w_{1,1}dw_{1,2}aw_{1,3}$.

A natural generalization of rw is that of introducing symbols (integers) showing which translation element is to be chosen. In our case the rw associated to the first production is:

$$w_{1,1}a_22w_{1,2}a_21w_{1,3}\,.$$

Naturally this *rw* will be written on the pushdown tape according to the conventions used for v_i' (see Section 4), i.e.:

$$c_2 1 a_2 \mid_1 c_1 2 a_2 \,.$$

At a certain point the machine will be in the following configuration:

$$[1, a_2, \mathrm{eq}_1] \quad Z_0 c_2 1 a_2 \mid_1 c_1 2 \quad [_1 b ; [_2 a]_2]_1 \,.$$

It will be enough, then, to reach our purpose, to add the following formulae to the transducer definition:

$[i, X, \mathrm{eq}_1], n, - \to (0, [i, X, n], E, \varnothing) \quad n \in N$

$[i, X, n], F, - \to (0, [+, n, \mathrm{start}], E$ and print $XZ_0, w_{i,k}$

if $F = c_k \in \{c_r\}$

$[+, n, \mathrm{start}], Z_0, [_i \to (1, [i, +, \mathrm{eq}], i\text{-rule}_n, \varnothing)$

the writing "i-rule$_n$" stands for the *rw* that defines the n-th translation element of the i-th production

$[+, n, \mathrm{start}], -, T \to \big(0, [+, +, b\text{-nest}_1], \varnothing, \mu_n(T)\big)$

the writing $\mu_n(T)$ stands for the value of terminal T computed using the n-th translation element.

We can, therefore, conclude that:

A pushdown two-way deterministic machine with output, as defined above, *is capable of implementing the GSDT schemes.*

The above machine is an alternative, in conventional automata theory, to the tree automata, introduced by Aho and Ullman [1] and by Thatcher [12], to implement GSDT's.

5.3. Extension of GSDT Schemes

The extension, we want to define, consists in simply extending the class of transformations for GSDTS, by including λ-operators. We obtain, in such a way, an useful and powerful tool in managing formal descriptions of block programming languages. GSDTS $+ \lambda$, in fact, allows us to distinguish the "meaning" of a substring according to the context in which the substring is situated.

Definition 5.2. A "*GSDTS* $+ \lambda$" is a generalization of a GSDTS. It will be denoted by a 7-tuple $V = (N, T, T', \Gamma, \Lambda, R, S)$, where N; T, T', Γ, S are defined as in Definition 5.1.

Λ is a finite set of λ-operators, λ_n, where n is an integer. R is a set of rules of one of the following forms:

1) $A \to \alpha,\ A_1 = \beta_1, \ldots, A_m = \beta_m$

2) $A \to abC,\ A_1 = \lambda_n \quad 1 \leqq n \leqq m$

subject to the following conditions:

1) The same three conditions of Definition 5.1,

2) $a, b \in T$ and $C \in N$.

We shall now define the *output* of a GSDTS $+ \lambda$. The value of A_i, when the translation element is of the form $A_i = \beta_i$ is computed as we have seen before. Suppose now that

$$A \to deF\,, \quad A_1 = \lambda_2$$

is a rule of a GSDTS + λ. Then, if the values of F_1 and F_2 at node F are y_1 and y_2, the value of A_1 is $y_2[d/e]$. (We mean with $x[a/b]$ the string obtained from x, replacing each occurrence of the symbol a by the symbol b). The subscript of λ, that is, indicates which translation of F must be used.

It is clear, now, that we can modify, in a manner analogous to the one used for the pdt, the *stack transducer* and make it *capable of implementing GSDTS* + λ.

We summarize the *relationships among translation schemes* in Fig. 3.

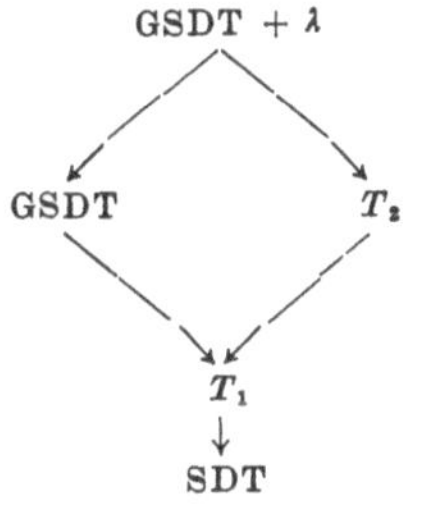

Fig. 3

The notation "$A \rightarrow B$", we see in Fig. 3, means that scheme B is included in scheme A.

Now we show in Table 1 the translation schemes and the machines which implement them.

Note that there is no equivalence between these machine and their respective schemes; the machines, that is, do not implement only translations defined by their respective scheme.

Table 1

SDT	1w pushdown transducer[1])
T_1	2w pushdown transducer
GSDT	2w pushdown transducer
T_2	stack transducer
GSDT + λ	stack transducer

[1]) 1w (2w) pushdown transducer means "one-way (two-way) deterministic pushdown automaton with output" and stack transducer means "two-way deterministic stack automaton with output".

Acknowledgement

The author is greatly indebted to his thesis advisor, Professor Luigi Petrone, for his generous interest and participation in this research.

References

[1] *Aho, A. V., J. D. Ullman,* Translations on a context-free grammar. Inform. and Control **19** (1971), 439–475.

[2] *Böhm, C., M. Dezani,* A CUCH-Machine: The automatic treatment of bound variables. Internat. J. Comp. Inform. Sc. (1971), 171–191.

[3] *Caracciolo di Forino, A.,* Some remarks on the syntax of symbolic programming languages. Comm. ACM **8** (1963), 456.

[4] *Church, A.,* The calculi of lambda-conversion. Ann. Math. Stud. No. 6, Princeton University Press, 1941.

[5] *Floyd, R. W.,* On the nonexistence of a phrase structure grammar for ALGOL 60. Comm. ACM **5** (1962), 483–484.

[6] *Ginsburg, S., S. A. Greibach, M. A. Harrison*, Stack automata and compiling. J. ACM **14** (1967), 172–201.
[7] *Hopcroft, J. E., J. D. Ullman*, Formal languages and their relation to automata. Addison-Wesley, Reading Mass. 1969; pp. 189–192.
[8] *Lewis, II, P. M. R. E. Stearns*, Syntax directed transduction. J. ACM **15** (1968), 464–488.
[9] *Naur, P.*, Report on the algorithmic language ALGOL 60. Comm. ACM **3** (1960), 299–314.
[10] *Petrone, L.*, Syntactic mappings of context-free languages. In: Proc. IFIP Congress, 1965; vol. 2; pp. 590–591.
[11] *Petrone, L.*, Syntax directed mappings of context-free languages. In: Conference Record of IEEE Ninth Annual Symposium on Switching and Automata Theory; 1968; pp. 160–175.
[12] *Thatcher, J. W.*, There's a lot more to finite automata theory than you would have thought. In: Proceedings of the Fourth Annual Princeton Conference on Information Sciences and Systems; 1970; pp. 263–276.

Abstract

A syntax-directed translation scheme, called T_2-semantics, is given. T_2-semantics, which assume a great usefulness in the compiling of block languages, are a generalization of T_1-semantics; they represent a more complete and realistic scheme for the complex process of translating programming languages. Furthermore, a stack transducer, which is capable of implementing T_2-semantics, is described and some meaningful examples of its computations are shown. Finally a similar extension of GSDT schemes is given to include processing of nested declarations; its implementation is carried out by means of a stack transducer.

Kurzfassung

Ein syntax-orientiertes Übersetzungsschema, T_2-Semantik genannt, wird vorgestellt. T_2-Semantiken, eine Verallgemeinerung von T_1-Semantiken, sind besonders geeignet für die Compilierung von Sprachen mit Blockstruktur. Sie stellen ein vollständigeres und realistischeres Schema für den komplexen Prozeß der Übersetzung von Programmiersprachen dar. Weiterhin wird ein Kellerautomat (stack transducer) beschrieben, mit dem T_2-Semantiken implementiert werden können. Einige Beispiele erläutern seine Arbeitsweise. Schließlich wird der Begriff des syntax-orientierten Übersetzungsschemas so verallgemeinert, daß er die Verarbeitung von geschachtelten Deklarationen einschließt; die Implementierung erfolgt wieder mittels eines Kellerautomaten.

Резюме

Приводится ориентированная на синтаксис модель перевода, так называемая T_2-семантика. T_2-семантики, имеющие большое значение для компиляции языков с блоками, являются обобщением T_1-семантик; они являются более полной и более реалистической моделью для сложного процесса перевода языков программирования. В дальнейшем описывается устройство перевода с помощью магазинной памяти, обеспечивающее имплементацию T_2-семантик, и приводятся несколько значительных примеров работы этого устройства. В окончание приводится подобное обобщение GSDT-модели с тем, чтобы вводить обработку последовательно вложенных декларации; ее имплементация осуществляется с помощью преобразователя с магазинной памятью.

(Received March 15, 1976)

Author's address:

M. Bert
Via S. Secondo, 19
10128 Torino
Italy

Elektronische Informationsverarbeitung und Kybernetik
EIK 13 (1977) 1/2, 97–106

Buchbesprechungen

Mathematical Foundations of Computer Science 1975. 4th Symposium, Mariánské Lázně, September 1—5, 1975. Edited by J. Bečvář. (Lecture Notes in Computer Science, 32). Springer-Verlag, Berlin-Heidelberg-New York 1975. 476 pp.

Der vorliegende 32. Band der von *G. Goos* und *J. Hartmanis* herausgegebenen Reihe „Lecture Notes in Computer Science" bietet den Konferenzbericht des vierten Symposiums „Mathematical Foundations of Computer Science", das vom 1. bis 5. September 1975 in Mariánské Lázně (ČSSR) stattfand. Veranstaltet wurde es im Rahmen der International Federation for Information Processing (IFIP) von dem Mathematischen Institut der Tschechoslowakischen Akademie der Wissenschaften zusammen mit mehreren tschechoslowakischen Institutionen und Gesellschaften technisch-naturwissenschaftlicher Richtung.

Der Band enthält die 12 einstündigen Hauptvorträge, die auf Einladung der Veranstalter gehalten wurden, sowie weitere 47 Kurzvortäge. Dem internationalen Charakter entsprechend waren Autoren aus zahlreichen Ländern vertreten. Die ČSSR stellte 15 Vorträge, 13 Vorträge wurden von Wissenschaftlern aus der UdSSR gehalten. Es folgten die BRD mit 8, Polen mit 6 und die DDR mit 5 Vorträgen. Vertreten waren weiterhin Italien (3), die USA (2), Großbritanien (2) sowie Dänemark, Frankreich, Rumänien und Ungarn (je 1).

Um die breite — man möchte sagen allzu breite — Zielsetzung der Veranstalter erkennen zu lassen, seien hier die auf Einladung gehaltenen Themen der Hauptvorträge angegeben. Eine Charakterisisierung der 47 Kurzvorträge übersteigt den vorliegenden Rahmen.

— Konstruktion eines vollständigen Mustersystems zur Prüfung auf Korrektheit (*Barzdin, Bičevskis, Kalninsh*)
— Zehn Jahre Beschleunigungsphänomen (Komplexitätstheorie) (*van Emde Boas*)
— Entdeckungs-Logik (künstliche Intelligenz) (Hájek)
— Über Sicherungsmodelle für Operationssysteme (*Harrison*)
— Zergliedern als Unteraufgabe beim Compilieren (*Král, Demner*)
— Parallel ablaufende rekursive Programm-Schemata (*Mazurkiewicz*)
— Über Probleme bei Pawlak-Maschinen (*Novotný*)
— Formale Potenzreihe und Wachstumsfunktionen von Lindenmayer-Systemen (*Salomaa*)
— Über die Darstellbarkeit von Relationen bei deterministischen und nichtdeterministischen Mehrband-Automaten (*Starke*)
— Über Probleme, die durch sukzessive Versuche lösbar sind (*Trachtenbrot*)
— Automaten und Kategorien (*Trnková*)
— Über Modelle zur Berechenbarkeit Boolescher Funktionen (*Zaslavskiĭ*)

W. Kämmerer

Beiträge zur Numerischen Mathematik. Herausgegeben von F. Kuhnert und J. W. Schmidt. Hefte 1 und 2. VEB Deutscher Verlag der Wissenschaften Berlin, R. Oldenbourg Verlag München-Wien, 1974. Heft 1: 236 S., Heft 2: 235 S.

Mit dem Erscheinen dieser Schriftenreihe wurde unter dankenswertem Einsatz der Herausgeber eine Möglichkeit zur Publikation speziell von Arbeiten aus dem Gebiet der Numerischen Mathematik für die DDR geschaffen und somit die große Bedeutung dieses relativ neuen Gebietes der Mathematik unterstrichen. In der Einleitung umreißen die Herausgeber das Ziel der „Beiträge" wie folgt: „Der numerische Algorithmus als zentraler Begriff der Numerischen Mathematik soll deshalb auch der inhaltlichen Gestaltung der Reihe zugrundeliegen. Dabei sollen alle Aspekte des numerischen Algorithmus, wie etwa seine Herkunft, Begründung, Konvergenz, Realisierung, Testung und Anwendung, im gleichen Maße in der Reihe erörtert werden können".

So findet man Beiträge von der Anwendung funktionalanalytischer Methoden in der konstruktiven Analysis bis hin zu konkreten Programmbeschreibungen und gewonnenen Rechenergebnissen. Die meisten Beiträge wurden von Mitarbeitern der TH Karl-Marx-Stadt bzw. der TH Dresden und den Berliner Numerikern (Zentralinstitut für Mathematik und Mechanik der AdW und Humboldt-Universität) beigesteuert.

Im 1. Heft findet der Leser die Untersuchung bestimmter Iterationsverfahren, wobei besonders Gleichungen der Art $F(x) = 0$ untersucht (*Burmeister, Kleinmichel*) bzw. Eigenwertprobleme mit dem Bazeley-Fox-Verfahren studiert werden (*Weinelt*). Die zweite Gruppe von Arbeiten beschäftigt sich mit dem Differenzenverfahren (*Schönheinz, Strehmel*), wobei z. T. sehr verbesserte Resultate gegenüber früheren Ergebnissen erreicht worden sind. Die Untersucheng von Approximations- und Interpolationsfragen beinhaltet die dritte Gruppe von Arbeiten, wobei u. a. Fragen der nichtlinearen Tschebyscheff-Approximation gelöst werden (*März*). Schließlich wird das Problem der Nullstellenbestimmung von Polynomen bzw. die Lösung von linearen Gleichungssystemen studiert, dabei werden konkrete Programmbeschreibungen vorgelegt (*Sandmann/Jahnke, Wallisch/Recknagel*). Das erste Heft enthält darüber hinaus einige Ergebnisse der konstruktiven Funktionentheorie, wie die Berechnung konformer Abbildungen (*Hofmann*), die potentialtheoretische Lösung eines ebenen Spannungsproblems (*Schinke/Grune*) sowie Untersuchungen über Einschließungsbereiche von Sattelpunktstrennkurven (*Thomas* †).

Im 2. Heft finden sich einige Arbeiten zur Behandlung von Eigenwertproblemen (*Jähnig, Lang, Lehmann*), wobei das Bazeley-Fox-Verfahren und die Grenzen seiner Anwendbarkeit, das Problem der Pseudostör-Iteration sowie Eigenwerte eines gestörten Operatorbüschels untersucht werden. Die zweite Gruppe von Arbeiten studiert die Lösung von Integralgleichungen (*Mach, Pohrath, Schneider, Schulz, Windisch*), dabei stehen Projektionsverfahren bzw. Quadraturformelmethoden im Vordergrund der Betrachtungen. Die durchgeführten Untersuchungen beziehen sich auf die Wiener-Hopf-Gleichung, auf Volterra-Gleichungen, die Behandlung nichtlinearer Schwingungsprobleme, die Lösung singulärer Integralgleichungen sowie auf einen speziellen Typ nichtlinearer Integralgleichungen. Wiederum sind eine Reihe von Arbeiten zur Behandlung von Gleichungssystemen, von algebraischen Gleichungen und zur Nullstellenbestimmung von Polynomen vorgelegt worden (*Ahlbrand, Bachmann, Fröhner, Mönch*). Es schließen sich Untersuchungen an zur Behandlung von Anfangs- und Randwertproblemen bei gewöhnlichen und partiellen Differentialgleichungen (Differenzenschemata für elliptische Gleichungen (*Komusiewicz*), Fehlerabschätzungen bei Anfangswertproblemen (*Oberländer*), Differenzenverfahren für hyperbolische Gleichungen (*Seifert*), Runge-Kutta-Nyström-Verfahren (*Weiss/Scholz*)). Schließlich findet der Leser Ausführungen über das Majorantenprinzip bei der Mehrschrittiteration (*Hoyer*) sowie Untersuchungen über ableitungsfreie Verfahren zur Extremwertbestimmung (*Hans*).

Die kurze Vorstellung der Arbeiten ist natürlich nicht ausreichend, ihre Bedeutung und Originalität voll zu würdigen. Es darf aber angenommen werden, daß den „Beiträgen" eine breite internationale Anerkennung nicht versagt bleiben wird. J. Nietzsch

D. Revuz: Markov Chains. (North-Holland Mathematical Library, Vol. 11). North-Holland Publishing Company Amsterdam-Oxford, American Elsevier Publishing Company New York, 1975. 336 pp.

Diese Monographie stellt eine moderne Einführung in die Theorie Markovscher Ketten, d. h. Markovscher Prozesse mit diskretem Zeitbereich und beliebigem Phasenraum, dar. Das Studium des Buches setzt umfangreiche Kenntnisse, u. a. von Maßtheorie, Algebra und Funktionalanalysis, voraus.

Die Definition der Markovschen Kette benutzt den Begriff der Zufallsgröße und vermeidet so den bisweilen etwas unanschaulichen Zugang, wie er z. B. von *Dynkin* eingeführt wurde. Der Autor achtet überall auf strengste Exaktheit, u. a. durch konsequente Benutzung der verschiedensten σ-Algebren. Dies wird z. B. an den Definitionen der Stopzeit und der strengen Markov-Eigenschaft deutlich.

Im weiteren werden nur homogene Markovsche Ketten betrachtet. Das Buch bietet einen ausführlichen Überblick über viele Eigenschaften dieser speziellen stochastischen Prozesse, wie z.B. Transienz und Rekurrenz, Ergodizität u. a. Sämtliche aufgeführten Aussagen werden streng bewiesen. Der Potentialtheorie speziell für „Harris-Ketten" wird ein eigener Abschnitt eingeräumt. An mehreren Stellen finden sich umfangreiche Resultate der Erneuerungstheorie. Diese sind allerdings — wie auch andere Ergebnisse — für den Anwender der Theorie Markovscher Ketten aufgrund ihrer teilweise abstrakten, nach größter Allgemeinheit strebenden Darstellung nicht immer leicht zugänglich.

Hervorzuheben sind die zahlreichen Übungsbeispiele am Ende eines jeden Abschnittes, welche das systematische Studium des Buches unterstützen. Diese Beispiele bereichern gleichzeitig die dargestellte Theorie um weitere interessante Aussagen. A. Neumann

Nasir Ahmed, K. Ramamohan Rao: Orthogonal Transforms for Digital Signal Processing. Springer-Verlag, Berlin-Heidelberg-New York 1975. 263 pp., 129 fig.

Das Buch behandelt die Möglichkeiten der Filterung und Kompression von Datenmassiven — wie z.B. Bildinhalte, medizinische Aufzeichnungen — mit Hilfe der Entwicklung nach orthogonalen Funktionssystemen unter besonderer Berücksichtigung der Anwendung von Digitalrechnern.

Nachdem in den ersten Kapiteln anhand der Fourier-Transformation die allgemeinen Grundlagen dargelegt werden, beginnen die spezifisch auf Rechnerbearbeitung bezogenen Teile mit der Einführung des schnellen Fourier-Algorithmus. Danach werden die digitalen Funktionen — Rademacher-, Haar- und Walsh-Funktionen — eingeführt, wobei besonders auf Letztere im Zusammenhang mit Hadamard-Matrizen in großer Ausführlichkeit eingegangen wird. Schnelle Algorithmen (in Analogie zur schnellen FT) werden sowohl für die Transformation von nach Hadamard geordneten Walsh-Funktionen als auch für die natürlich geordneten Walsh-Funktionen entwickelt. In einem weiteren Kapitel werden einige Transformationen vorgestellt, welche nicht so allgemein bekannt sind, so z. B. eine verallgemeinerte Transformation, welche einen Übergang zwischen der Walsh-Transformation und der Fourier-Transformation ermöglicht, wobei besonders der Spezialfall einer komplexen Walsh-Transformation von großem Interesse ist, weiterhin die diskrete Cosinus-Transformation, deren Grundvektoren große Ähnlichkeit mit den Eigenvektoren des Markov-Prozesses 1. Ordnung haben und die deshalb in vielen Fällen zu sehr schnell konvergierenden Approximationen führt.

Die Anwendung der Methoden auf die digitale Signalbehandlung wird ab Kap. 8 behandelt, und zwar folgende Fälle: Wiener-Filter mit minimalem Aufwand (im Idealfall als Diagonalmatrix), Datenkompression von EKG-Daten und von Bildern. Obwohl die Fälle theoretisch auf eine Karhunen-Loeve-Entwicklung führen, die sich natürlich i. allg. nicht auf schnelle Algorithmen hinführen läßt, erweist es sich interessanterweise, daß eine Reihe von leicht realisierbaren Transformationen, wie die diskrete Cosinus-Transformation u. a., zu nahezu optimalen Ergebnissen führen und sogar die Walsh-Transformation noch recht gut verwendbar ist. Einige sehr praktische Auswahlkriterien für die Eignung verschiedener Transformationen werden hier entwickelt. Im letzen Kapitel werden schließlich einige interessante Gedanken zur Figurenerkennung und Klassifikation vorgestellt.

Das Buch kann allen, die sich mit Problemen wie Kompression, Filterung und Klassifikation von Bildinhalten und Kurvenverläufen auf Digitalrechnern beschäftigen, empfohlen werden. Es bietet einen vollständigen Überblick über die wesentlichen Algorithmen und darüber hinaus eine Fülle von Anregungen. K.-H. Schmelovsky

Helmuth Späth: Cluster-Analyse-Algorithmen zur Objektklassifizierung und Datenreduktion. (Reihe: Verfahren der Datenverarbeitung). R. Oldenbourg Verlag, München-Wien 1975. 217 S., 48 Abb., 13 Tab.

Innerhalb der Klasse der multivariaten Verfahren erweisen sich die Clusterverfahren in fast allen empirischen Wissenschaften bei der Aufstellung von Klassifikationsschemata zum Zwecke der Datenreduktion oder zur Unterstützung der Hypothesenbildung als bedeutsam. Die allgemeine Problemstellung besteht darin, eine Menge X von Objekten auf

Grund einer Beschreibung der Objekte durch Werte mehrerer Variabler oder auf Grund gegebener Abstands- oder Ähnlichkeitsbeziehungen zwischen Objekten in eine Menge von Teilen mit vorgegebenen Eigenschaften zu zerlegen, Die Clusterverfahren unterscheiden sich gerade in den Festlegungen eines entsprechenden Zielkriteriums und der algorithmischen Vorgehensweise.

Im vorliegenden Buch werden einige in der Praxis bewährte Verfahren zur Erzeugung einer Zerlegung der Objektmenge X bzw. einer Folge von Zerlegungen vorgestellt. In Kap. 1 werden als Voraussetzungen für die Anwendung von Clusterverfahren Betrachtungen über Abstands- bzw. Ähnlichkeitsfunktionen für Objektpaare und Paare von Objektmengen dargestellt, wobei sowohl metrische als auch ordinale, nominale und gemischte Daten zum Ausgangspunkt genommen werden. Das Kapitel über partitionierende Clusteralgorithmen enthält einfache Verfahren, die eine Zerlegung operational definieren und oft als Vorstufen für aufwendigere Verfahren eingesetzt werden. Den Hauptteil bilden Beschreibungen heuristischer Algorithmen, die auf die Minimierung der Summe der innerhalb von allen Clustern gebildeten Summen der quadrierten Abstände der Clusterelemente von ihrem Schwerpunkt gerichtet sind. Das Grundprinzip ist die schrittweise Verbesserung einer Zerlegung im Hinblick auf die Zielfunktion durch Verschiebung von Elementen zwischen Clustern. Varianten dieser Vorgehensweise für andere Zielfunktionen und mit anderen Grundoperationen werden angegeben. Neben den Beispielen, an denen die Demonstration der Vorgehensweise und der Problematik der Clusteranalyse erfolgt, wird besonders das multiple Standortproblem behandelt.

Ein weiteres Kapitel enthält die wichtigsten hierarchischen Clusterverfahren, welche spezielle Folgen von Zerlegungen erzeugen. Die dabei verwendeten Grundoperationen sind die Fusion bestehender Cluster und die Aufspaltung eines Clusters. Die entsprechenden auf Agglomeration bzw. Zerlegung beruhenden Verfahren unterscheiden sich hinsichtlich der Kriterien zur Auswahl der zu fusionierenden bzw. aufzuspaltenden Cluster. Den Abschluß bilden einige spezielle Verfahren wie z.B. die Umordnung einer Ähnlichkeitsmatrix in Blockform.

Das Buch ist vorrangig auf die Darstellung von Clusterverfahren gerichtet, die sich in der Praxis bewährt haben, obwohl teilweise ihre mathematische Fundierung fehlt. Bezüglich einer umfassenderen Sicht der mathematischen Grundlagen der Clusteranalyse sowie des Entwicklungsstandes der Verfahren sei auf den Band „Automatische Klassifikation. Theoretische und praktische Methoden der Gruppierung und Strukturierung von Daten" von *Bock* verwiesen.

Der besondere Wert des vorliegenden Buches liegt in der Darstellung praktikabler Algorithmen mit den dazugehörigen FORTRAN-Programmen sowie in den über die Beispiele gegebenen Hinweisen für die mit der Anwendung verbundenen Probleme wie Wahl der geeigneten Zielkriterien und Algorithmen, Auswahl der Objektbeschreibung und Festlegung der Clusteranzahl als auch in der Demonstration von Strategien zur Behandlung von Clusterproblemen. H. Hagendorf

Richard B. Kieburtz: Structured Programming and Problem Solving with ALGOL W. Prentice-Hall, Inc. Englewood Cliffs, New Jersey 1975. 365 pp.

Abweichend von den meisten Lehrbüchern über das Programmieren in einer problemorientierten Sprache behandelt der Autor schwerpunktmäßig ein systematisches Herangehen an den Entwurf von Algorithmen und lediglich ergänzend die Umsetzung der Notation eines Algorithmus in eine problemorientierte Programmiersprache. Das Bedürfnis nach einer Unterweisung in einer Entwurfsmethodik empfindet der Studierende allerdings kaum bei den (notwendigerweise) relativ einfachen Beispielen eines einführenden Lehrgangs, sondern erst später bei den berufsbedingten Aufgaben und Problemen, die nur noch kollektiv zu bewältigen sind. Genau auf diese Phase will das vorliegende Buch den Leser durch Erläuterung von Verfahrensweisen vorbereiten, die seit einigen Jahren in der internationalen Literatur unter dem Schlagwort „Strukturiertes Programmieren" bekannt geworden sind.

Aus dieser Zielstellung ergibt sich folgerichtig die Entscheidung des Autors für eine neuartige graphische Darstellungsform, die das Strukturieren eines Algorithmus durch seine

Anschaulichkeit und innere Logik zwingend unterstützt. Der Algorithmus wird mit Hilfe von sog. Struktogrammen (Nassi-Shneidermann-Diagrammen) notiert. Diese Notation ersetzt nicht nur vollwertig die bisher eingebürgerte Symbolik der Programmablaufpläne, sondern hat den Vorzug, eine zwangsweise Führung der gedanklichen Arbeit zu übernehmen und damit den strukturierten Entwurf selbst entscheidend zu fördern.

Nicht so zwingend ist die Entscheidung für eine problemorientierte Programmiersprache als eine rechnerverständliche Notation eines Algorithmus. Hier fiel die Wahl auf die an der Stanford University, California, USA entwickelte Sprache ALGOL W, die der methodischen Zielsetzung des Autors sehr entgegenkommt. Es handelt sich bei ALGOL W um eine erweiterte Variante von ALGOL 60. Beispielsweise sind als Erweiterungen aufgenommen: eine CASE-Anweisung für den Fall einer nichtbinären Verzweigung (etwa ähnlich dem computed GOTO in FORTRAN) und eine RECORD-Vereinbarung für den Fall einer hinsichtlich Datentyp heterogenen Datenstruktur (etwa ähnlich der Struktur-Vereinbarung in PL/I). Beide Erweiterungen werden in Hinblick auf den einführenden Charakter in diesem Buch nicht behandelt.

Eine typische Vorgehensweise beim „Strukturierten Programmieren" ist das schrittweise Untersetzen der einzelnen Anweisungen eines Algorithmus. Das spiegelt sich in den schließlich vorliegenden Programmen darin wider, daß sie reichlich Kommentare enthalten; dadurch ist die Lesbarkeit der Programme wesentlich verbessert. Sozusagen als Krönung der Unterweisung in der Technik der schrittweisen Untersetzung wird für die bekannte Denksportaufgabe über acht Königinnen auf dem Schachbrett ein algorithmicher Entwurf vorgeführt.

Allen, die nicht irgendeine problemorientierte Programmiersprache erlernen möchten, sondern anstreben, eine systematische Entwurfsmethodik für Algorithmen zu beherrschen, muß das Studium des besprochenen Buches dringend empfohlen werden. W. Feitscher

Karl-Heinz Werler: Probleme der graphischen Datenverarbeitung. (Elektronisches Rechnen und Regeln, Band 7). Akademie-Verlag, Berlin 1975. 270 S., 224 Abb.

Das Buch wendet sich in erster Linie an die künftigen Anwender der grafischen Datenverarbeitung (GDV). Es vermittelt die besonderen Techniken der GDV ohne mehr als grundlegende Kenntnisse der EDV vorauszusetzen.

Der Stoff wurde in 3 Kapitel unterteilt. In Kap. 1 „Allgemeine theoretische Grundlagen" (71 S.) werden neben der Abgrenzung des Gegenstandes die mathematischen Hilfsmittel in Erinnerung gebracht (homogene Koordinaten, Projektionen) und spezielle von der GDV angewendete rechentechnische Methoden (Listen-, Ringstrukturen, hierarchische Beschreibungen, grafische Programmiersprachen) erläutert. Ausführlich wird auf die interaktive Arbeitsweise eingegangen.

Im umfangreichsten Kapitel 2 „Rechentechnische Grundlagen" (123 S.) werden Geräte- und Programmtechnik behandelt. Der technische Aufbau von Ein- und Ausgabegeräten für Zeichnungen (Digitalisiergeräte, Zeichenmaschine, Plotter, Mikrofilmausgabe) und 3-dimensionale Körper wird beschrieben. Sehr ausführlich (44 S.) werden die Eigenschaften grafischer Bildschirmgeräte dargestellt. Dabei werden jeweils zunächst die allgemeinen Prinzipien und dann typische Geräte erläutert. Im Abschnitt über Programmiertechnik werden die Besonderheiten der Ein- und Ausgabe bei der GDV betrachtet. Ausführlich wird wiederum auf die Programmierung von interaktiven Geräten eingegangen.

Kap. 3 „Anwendungen der GDV" (54 S.) beabsichtigt die zielgerichtete Diskussion der Einsatz- und Anwendungsmöglichkeiten. Dabei werden hauptsächlich die grafisch-geometrischen Aspekte bei Anwendungen in der technischen Vorbereitung der Produktion betrachtet. Es werden anhand von 8 charakteristischen Beispielen die Techniken der interaktiven grafischen Arbeitsweise bei bestimmten Anwendungsgebieten dargestellt (z.B. Leiterplattenentwurf, Entwurf von integrierten Schaltungen, Aufbau und Manipulation 3-dimensionaler Gebilde, Programmierung von NC-Maschinen).

Ein Literaturverzeichnis mit 398 Literaturstellen erlaubt dem interessierten Leser die weiterführende Literatur zu erschließen.

Das vorliegende Buch gibt mit seinen beiden ersten Kapiteln eine gute Einführung in das Gebiet der GDV. Es faßt die in einer Vielzahl von Publikationen verstreuten Informationen unter Vernachlässigung der unwesentlichen Details geschickt zusammen. Überflüssig ist höchstens die Wiedergabe der Grundlagen konventioneller Geräte- und Programmtechnik; anstelle von 2.2.1 und 2.3.1. wäre ein Anhang mit Begriffsbestimmungen zweckdienlicher. Eine Straffung des Buches ist jedoch durch Verringerung der im Zusammenhang mit der GDV betrachteten Problemkreise (interaktive Arbeitsweise, Mensch-Maschine-Kommunikation, rechnergestützte Arbeitsweise, Dialog, Konstruktionstheorie, Problemlösungsprozesse usw.) möglich. Damit würde Platz für die Definition grundlegender Begriffe geschaffen (z.B. bleibt offen, was denn eigentlich GDV ist). Die Diskussion der im Titel des Buches genannten Probleme der GDV wurde vermißt.

Der Versuch, Klassen von Anwendungen auf der Grundlage typischer Arbeitsweisen zu formulieren, führt zum Verlust an Detailliertheit. Für den anvisierten Leserkreis des Buches, die zukünftigen Anwender der GDV, bleibt damit verborgen, welche Probleme die Projektierung und Anwendung der GDV mit sich bringt. Die Beschränkung auf wenige ausführlich beschriebene Einsatzfälle wäre nach Ansicht des Rezensenten zweckdienlicher gewesen.

Nicht verständlich ist, weshalb Probleme der Ökonomie und Effektivität bewußt ausgeklammert werden. Für den Rezensenten bleibt somit für die von ihm einschätzbaren Anwendungsfälle die Frage offen, ob durch die GDV ökonomisch vertretbare Lösungen erreicht werden können. Beim Gebiet der Rationalisierung und Automatisierung der technischen Vorbereitung der Produktion, welches „in den Mittelpunkt der Betrachtungen und Beziehungen zur Praxis“ (S. 1) gestellt wurde, ist der Nutzensnachweis eines EDV-Projektes jedoch entscheidend.

Ungeachtet dieser Einschränkungen stellt das Buch einen notwendigen und nützlichen Beitrag zur EDV-Literatur dar. Ihm ist eine gute Aufnahme zu wünschen. L. Koch

Biokybernetik, Band V. Materialien des IV. Internationalen Symposiums „Biokybernetik“, Leipzig, 19.—22. September 1973. Herausgeber: H. Drischel und P. Dettmar. VEB Gustav Fischer Verlag, Jena 1975. 368 S.

Der 5. Band der bewährten Reihe gibt den größten Teil der auf dem im Titel genannten Symposium gehaltenen Vorträge wieder. Es ist unmöglich, die 47 hier abgedruckten Vorträge im einzelnen zu referieren. Außer dem Beitrag von *Kämmerer* über algebraische und graphentheoretische Strukturen im Rahmen der Kybernetik, der als Einführung steht, wurden die übrigen zu folgenden Themen-Komplexen zusammengefaßt: Informationsverarbeitung im Zentralnervensystem, Rezeptormechanismen, Informationsverarbeitung im visuellen System, Mustererkennung im visuellen System, Mustererkennung (allgemein und technisch) sowie Regelung und Adaptation.

Dieser Band zeigt, verglichen mit dem ersten Band der Reihe, den eindrucksvollen Fortschritt auf diesem Gebiet und den hohen heuristischen Wert einer theoretisch gut fundierten Biokybernetik. Der Schwerpunkt liegt im visuellen System, wobei auch teilweise stark divergierende Auffassungen vertreten sind. Biologen, Physiologen, Mediziner, Techniker, Mathematiker und Systemtheoretiker kommen hier zu Wort; der multidisziplinäre Ansatz der Biokybernetik wird in seiner hohen Valenz erkennbar. Der vorzüglich ausgestattete Band belegt dies eindrucksvoll.

Man mag bedauern, daß die teilweise lebhaften Diskussionen auf diesem Symposium, an dem etwa 180 Wissenschaftler teilnahmen, nicht übernommen werden konnten. Den Herausgebern ist für mühevolle Redaktionsarbeit zu danken, der Band setzt als gewichtiges Glied die Biokybernetik-Reihe fort, die in einer naturwissenschaftlich orientierten Bibliothek nicht fehlen sollte; Kybernetiker, Physiologen, Mediziner, Biologen und Psychologen können hier vielfältige Anregungen und Informationen empfangen. G. Tembrock

Computers in Education. Proceedings of the IFIP 2nd World Conference, Marseilles, France, September 1—5, 1975. Edited by O. Lecarme and R. Lewis. North-Holland Publ. Comp. Amsterdam-Oxford, American Elsevier Publ. Comp. New York, 1975. 1020 pp.

Es handelt sich hier um das wesentlich gekürzte Protokoll der von der „Internationalen Föderation für Informationsprozesse" organisierten 2. Weltkonferenz zur Problematik der Computer-Anwendung in der Bildung, die vom 1. bis 5. September 1975 in Marseilles stattfand. Die in 70 Sektionen gegliederte Konferenz behandelte diese Problematik unter vielfältigsten Aspekten aus der Sicht aller Bildungsbereiche. Neben Vertretern aus kapitalistischen Staaten und Entwicklungsländern nahmen auch Wissenschaftler aus der UdSSR, der DDR und anderen sozialistischen Staaten teil, so daß der Protokollband einen repräsentativen Überblick über den internationalen Entwicklungsstand der Computer-Anwendung im Bildungswesen vermitteln dürfte.

Allein die Nennung der übergreifenden Themenstellungen der Sektionen — von den nahezu 300 Einzelbeiträgen nicht zu reden — würde den Rahmen einer solchen Kurzbesprechung sprengen. Daher kann nur ein Überblick zu wesentlichen Inhaltsbereichen der Konferenz gegeben werden.

Übergreifende theoretische Fragen der CMI (computergesteuerte Instruktion) und CAI (computerunterstützte Instruktion) machen einen im Vergleich zur Behandlung der Anwendungsproblematik geringeren Anteil der Gesamtaussagen der Konferenz aus. Der Band enthält einige Beiträge zur Anwendung verschiedener Programmsprachen, zur Steuerung algorithmischer und heuristischer Prozesse sowie zur Mensch–Maschine-Problematik.

Im Bereich der allgemeinbildenden Schule werden Computer-Anwendungen zur Bildungsorganisation und -planung, bei verschiedenen Projekten der CAI und CMI im mathematisch-naturwissenschaftlichen Unterricht sowie zur Lehrerausbildung beschrieben.

Einen größeren Umfang nehmen Berichte über Computer-Anwendungen in der Hoch- und Fachschulbildung ein. Das Spektrum der Aussagen reicht hier von der Curriculumentwicklung über die Lehrstoffvermittlung in Mathematik, Physik, Chemie, Biologie, Fremdsprachen, Ökonomie, Soziologie, Medizin, den Ingenieurwissenschaften bis hin zur computerunterstützten Organisation der Dokumentation und Information.

Auf besonderes Interesse dürften die Berichte über die vielfältigen Potenzen der Computernutzung als Hilfsmittel speziell der mathematischen, ökonomischen, soziologischen und medizinischen Forschung stoßen.

Breiten Raum nehmen auch Aussagen über die Computeranwendung für Bildungszwecke in Management, Industrie und Handel ein. Als spezielle Anwendungsgebiete seien weiterhin die Testkonstruktion, die Operationsforschung sowie die Kunstwissenschaften (Musik, bildende Künste) hervorgehoben.

Schließlich befaßte man sich in einigen Sektionen mit ökonomischen und technologischen Aspekten der Computeranwendung im Bildungswesen sowie mit Erfahrungen, die mit CMI und CAI in Entwicklungsländern gewonnen wurden.

Insgesamt liegt mit diesem Buch ein Informationsspeicher vor, der auf Grund seiner außerordentlich kurzen Entwicklungszeit hochaktuell ist und dessen Wert für Forschung und Praxis in den verschiedenen Anwendungsgebieten kaum überschätzt werden kann.

K.-H. Walter

Wolfgang Hilberg: Elektronische digitale Speicher. (Reihe: Datenverarbeitung). R. Oldenbourg Verlag, München-Wien 1975. 260 S., 480 Abb., 26 Tab.

Genau genommen müßte der Titel des Buches Ferritkern- und Halbleiterspeicher lauten, denn hierauf ist im wesentlichen sein Inhalt beschränkt. Der Verfasser beginnt allerdings mit einer fast 70 Seiten umfassenden allgemeinen und interessanten Einführung. Hierin werden das Verhältnis Mensch—EDVA vom Standpunkt der Speicher deutlich herausgearbeitet und Konsequenzen für die Speicherhierarchien gezogen. Dabei werden weiter Folgerungen für den digitalen Speicher, für die Kapazität zur Adressierung, optimale Modulkapazität bis zum assoziativen Speicher und Ausnutzung der CPU abgeleitet. Etwa 100 Seiten sind dem Ferritkernspeicher gewidmet. Hier werden neben vielen Details vor allem prinzipielle Funktion, Organisationsformen (3-, 2- und $2^1/_2$-D), dynamische Eigenschaften, elektronische Schaltungen wie Schreib- und Leseverstärker, Kostenfragen und Zuverlässigkeit behandelt. Neu dürfte sein, daß für magnetische Betrachtungen das Preisach-Diagramm mit benutzt wird. Ca. 130 S. sind den verschiedenen Formen der Halbleiterspeicher gewidmet. Sie beginnen mit der Wirkung des Flip-Flop, wobei u. a. Frequenzgrenzen, Laufzeit, Stabilität, Triggervorgang und Schaltungsvarianten: DCTL, TTL, ECL sowie Sonderformen bei großen Speichern in Form von typischen Speicherzellen und Organisationsfragen betrachtet werden. Ein kurzer Abschnitt geht betont auf das Lay-out und Minimalflächen je bit ein.

Neben den Speichern mit der Grundeinheit des Flip-Flop werden auch die dynamischen Speicher in ihrer Vielfalt kurz behandelt. Auffallend ist dabei, daß die Schieberegister von den Varianten der CCD- bzw. BBD-Speicher durch die Festwert- und Assoziativspeicher in der Reihenfolge getrennt sind und daß hier auch die bubble-memory einbezogen werden. Den Abschluß des Buches bilden einige Betrachtungen zu Fertigungsgesichtspunkten bei Halbleiterspeichern, wie ein mathematisches Modell der Ausbeute, Fragen der Kosten und Herstellungstoleranzen. Ferner sind ein Literaturverzeichnis und ein guter Bildanhang über Strukturen von hochintegrierten Schaltkreisen vorhanden.

Insgesamt weicht das Buch von der Mehrzahl der jetzt vielfältig erscheinenden Speicherbücher ab. Es enthält stellenweise viel eigenständiges neuartiges Material, ist hier dann also im Sinne einer guten Monographie vertieft. An anderen Stellen ist es aber gut allgemeinverständlich geschrieben. Hierdurch wirkt es auf den Leser etwas inhomogen. Der oft etwas subjektiv vom Autor gewählte Standpunkt wirkt aber meist anregend, was insbesondere bei der schnellen Entwicklung des Halbleitergebietes wohl begrüßt werden kann.

H. Völz

Samiel H. Fuller: Analysis of Drum and Disk Storage Units. (Lecture Notes in Computer Science, 31). Springer-Verlag, Berlin-Heidelberg-New York 1975. 283 pp.

Der Inhalt dieser Monographie ist weitaus spezieller als sein Titel vermuten läßt. Er betrifft lediglich theoretische und experimentelle Untersuchungen zur Zugriffszeit bei Trommeln und Platten mit festen und bewegten Köpfen. Sie geht auf eine Dissertation des Autors von 1972 zurück. Es werden folgende Betriebsarten untersucht:

FIFO (first-in-first-out)
SLFT (Shortest-latency-time-first)
MTPT (minimal-total-processing-time).

Die umfangreichen, oft neuen theoretischen Ansätze und Ergebnisse geben Hinweise zur Erhöhung der Effektivität in der Zusammenarbeit von Zentraleinheit des Rechners und Speicher. Viele Kurven (Rechenergebnisse) veranschaulichen die Zusammenhänge. Experimentell gewonnene Ausdrücke verbinden Theorie und Praxis.

Für Interessenten dieses schmalen, aber für die Effektivität der Rechentechnik bedeutsamen Gebietes dürfte die Monographie erhebliche Bedeutung besitzen.

H. Völz

Karl Hormann: Direkte Methoden der Stabilitätsprüfung. VEB Verlag Technik Berlin 1975. 206 S., 93 Abb., 3 Tafeln

Der vorliegende Band aus der Reihe „Theoretische Grundlagen der technischen Kybernetik" stellt sich das Ziel, Verfahren der Stabilitätsanalyse dynamischer Systeme, die auf der Frequenzkennlinien- bzw. der Zustandsraummethode beruhen, einem breiten Leserkreis zugänglich zu machen. Algebraische Methoden werden nicht berücksichtigt, weil sie bereits allgemein in die Ingenieurausbildung aufgenommen worden sind und deshalb als bekannt vorausgesetzt werden können.

Zum Verständnis des Buches werden die Grundlagen der komplexen Funktionentheorie, der Operatorenrechnung und der Vektoralgebra vorausgesetzt. Da das Buch vorrangig Ingenieuren als Hilfsmittel dienen soll, sind die mathematischen Sachverhalte weitestgehend mit dem regelungstheoretischen Begriffsapparat beschrieben. Es ist dem Autor bis auf wenige Ausnahmen gut gelungen, dabei die größtmögliche Strenge und Exaktheit der Formulierungen zu wahren. Auf Beweise wird verzichtet, wenn die Aussagen sich anschaulich erklären lassen oder der Beweisapparat die in der Ingenieurausbildung vermittelten Kenntnisse übersteigt. Grundsätzlich wird die Erörterung der Sachverhalte durch praktische Beispiele wirkungsvoll unterstützt.

In einem einleitenden Abschnitt werden die Begriffe dargelegt, die für Stabilitätsuntersuchungen gewöhnlicher Differentialgleichungssysteme benötigt werden. Die nächsten vier Abschnitte sind Verfahren und Kriterien für lineare Systeme mit konstanten Parametern gewidmet. Zunächst werden aus dem Residuensatz der Funktionentheorie das Nyquist-Kriterium, das Zweiortskurvenkriterium und das Umlaufkriterium abgeleitet, dann wird eine Methode angegeben, für zwei freie Systemparameter die Stabilitätsgebiete zu ermitteln. In den letzen vier Abschnitten werden nichtlineare Stabilitätsprobleme behandelt. Dabei wird zunächst im Fall stetiger Nichtlinearitäten Ljapunows Satz über die Stabilität der linearen Näherung erläutert und dann die Methode der Beschreibungsfunktion zur Untersuchung der Stabilität von Eigenschwingungen im Falle unstetiger Nichtlinearitäten dargelegt. Für Differentialgleichungen zweiter Ordnung werden Phasenporträts untersucht; schließlich wird die direkte Methode von Ljapunow anschaulich erläutert und das Popov-Kriterium mit einigen Verallgemeinerungen daraus abgeleitet.

Im Literaturverzeichnis werden die zitierten Arbeiten und die wichtigsten Werke aus der ingenieurtechnischen Fachliteratur über Stabilitätsfragen zusammengestellt; vermißt werden einige internationale theoretische Standardwerke (*Demidovič*, *Halanay*, *Rozeau*, *Rojtenberg*, *Zubov* u. a.). E. Griepentrog

A. H. Bobeck, E. Della Torre: Magnetic Bubbles. (Series of Monographs on Selected Topics in Solid State Physics, 14). North-Holland Publ. Comp., Amsterdam-Oxford; American Elsevier Publ. Comp., Inc., New York; 1975. 222 pp.

In den letzten Jahren hat die magnetische Blase (bubble) als eine spezielle Eigenschaft magnetischer Schichten große Anwendungsrelevanz erlangt. Mit ihr besteht die Möglichkeit, digitale Speicher zu realisieren, die schneller als Plattenspeicher sind und deren Kapazität größer als bei Halbleiter- und Kernspeichern sein könnte. In dieser Hinsicht würden sie eine echte Lücke füllen. Ferner gestattet ihre Anwendung neben der Speicherfunktion auch die Realisierung von logischen Verknüpfungen. Das vorliegende Buch versucht, die wesentlichen Ergebnisse etwa vom Stand 1973 einheitlich darzustellen. Es beginnt mit einer anschaulichen Beschreibung der physikalischen Mechanismen, um dann zur Behandlung der theoretischen Beschreibung überzugehen. Gerade die geschlossene und größtenteils tiefgründige Darstellung aller wesentlichen Zusammenhänge ist gegenüber dem Originalschrifttum sehr vorteilhaft. Es werden behandelt: das Stabilitätsgebiet mit Abgrenzungen der Streifendomänen und dem Collaps der Bubbles, die Dynamik der Bubbles, Einflüsse von Temperatur, Schichtparametern und Feldgeometrie u. a. bei *T–I*- und *Y–I*-bars. An Materialien werden vor allem die Orthoferrite und uniaxialen Garnets ausführlich behandelt. Bezüglich der Garnetfilme sind auch Herstellungsmethoden beschrieben. Schließlich erfolgt eine Behandlung der Ssyteme, u. a. Angelfish (dreieckförmig), spezielle Leiterschleifen, PERMALLOY-Punkte, *T–I*-, *Y–I*- und *V*-bars. Weitere Gebiete sind die Erzeugung sowie Vernichtung und Wiedergabe von bubbles(bits).

Insgesamt liegt damit ein kleines Handbuch für dieses Gebiet vor. Es hat sowohl bezüglich der Einführung in dieses Gebiet als auch im gewissen Sinne als kurze Monographie über dieses Gebiet seine Bedeutung. Das reichliche Bild- und Tabellenmaterial vermittelt die wesentlichen Fakten und veranschaulicht den Aufbau von Speichern. Ein umfangreiches Literaturvezeichnis (ca. 150 Zitate) gibt Hinweise auf weitere Details. Sachwort- und Namensverzeichnis erleichtern den Gebrauch. H. Völz

Tagungen

Das Zentrum für Rechentechnik der Akademie der Wissenschaften der DDR organisierte im Rahmen des Koordinierungsrates für Informationsverarbeitung (KORIV) Berlin und der Hauptforschungsrichtung Informationsverarbeitungssysteme eine

Sommerschule

„Algebraische Methoden bei der theoretischen Fundierung der Programmierung“ vom 25. bis 28. Mai 1976 in Bad Stuer (Plauer See).

Zielstellung der Veranstaltung war es, den Teilnehmern einen Überblick über den Stand der theoretischen Grundlagenforschung, insbesondere zur Anwendung der Algebra für die Fundierung der Programmierung zu geben. Die Kollegen Dr. Reichel, Dr. Kaphengst und Dr. Hupbach, die im VEB Robotron, Zentrum für Forschung und Technik seit einigen Jahren intensiv an dieser Problematik arbeiten, hielten Vorlesungen zu den Komplexen:

(1) Der „Fixpunkt-Ansatz“ von Dana Scott und dessen Weiterentwicklung als mathematische Theorie zur Behandlung von Problemen der Programmverifikation unter Einsatz von Mitteln der Verbandstheorie (*Hupbach*)
(2) Die algorithmischen Algebren von Gluschkow zur formalen Beschreibung von Maschinenprogrammen und Transformationen von Maschinenprogrammen (*Kaphengst*)
(3) Die algebraische Algorithmentheorie als Mittel zur Untersuchung rekursiver Algorithmen, rekursiver Datenstrukturen und der Semantik von Programmiersprachen (*Kaphengst*)
(4) Die Anwendung der Kategorientheorie zur Untersuchung von Flußdiagrammen (*Reichel*)

Bei der abschließenden Diskussion dominierte die Problemstellung, auf welche Weise eine geeignete Verbindung zwischen Theorie und Praxis geknüpft werden kann. Als wesentlich wurde die intensive Kommunikation zwischen Theoretikern und Praktikern erkannt. Die Sommerschule war dazu ein erster guter Schritt. Daneben kommt es darauf an, daß in verstärktem Maße die Rechentechniker bereitstehende mathematische Methoden für die Untersuchung und Fundierung programmiertechnischer Instrumente und Technologien nutzen und umgekehrt die Erarbeitung von nutzbaren mathematischen Methoden stimulieren.

Die Auffassungen vom praktischen Wert der theoretischen Untersuchungen schwankten sehr stark. Einer zu hohen Wertung der mathematischen Begründung wurde entgegengehalten, daß gerade in der Rechentechnik eine Reihe von ingenieurtechnischen, ökonomischen und psychologischen Aspekten sowie historisch entstandene Gegebenheiten eine Rolle spielen, die die Bereitstellung praktikabler Resultate oft wesentlicher beeinflussen als mathematische Theorien. Diese Situation muß stets berücksichtigt und es darf nicht nur vom theoretischen Wunschdenken ausgegangen werden.

Andererseits kann eine Vernachlässigung der theoretischen Erkenntnisse unabsehbare Auswirkungen mit sich bringen, die oft nur schwer korrigierbar sind. Ein zu pragmatisches Herangehen birgt die Gefahren der Unvollständigkeit und Widersprüchlichkeit in sich, die dann historische Gegebenheiten bestimmen.

Als gegenwärtig wichtige Arbeitsgebiete für eine theoretische Fundierung wurden

— Datenstrukturen und
— Kompliziertheit und Effektivität von Programmen

in der Diskussion herausgearbeitet. Einem dieser Problemkreise sollte sich die Sommerschule 1977 widmen.

P. Bachmann

FSC
www.fsc.org
MIX
Papier | Fördert
gute Waldnutzung
FSC® C083411

Kolibri 360 GmbH
Ferdinand-Jühlke-Straße 7
99095 Erfurt, Deutschland
produktsicherheit@kolibri360.de